# Lecture Notes in Mobility

**Series Editor**

Gereon Meyer, VDI/VDE Innovation + Technik GmbH, Berlin, Germany

Lecture Notes in Mobility (LNMOB) is a book series that reports on the latest advances in research, development and innovations for the intelligent, connected and sustainable transportation systems of the future, comprising e.g.:

- Electric and hybrid vehicles
- Energy-efficient vehicles
- Alternative and optimized powertrains
- Vehicle automation and driver assistance
- Clean and intelligent transportation systems
- Interfaces with transportation networks and power grids
- Mobility services
- Business models
- Public policy

LNMOB publishes edited conference proceedings, contributed volumes and monographs that present firsthand information on cutting-edge research results and pioneering innovations as well as new perspectives on classical fields, while maintaining Springer's high standards of excellence. Also considered for publication are dissertations and other related material of exceptionally high quality and interest. The subject matter should be original and timely, reporting on the latest developments in all areas covered by this series. Concerns about climate change, air quality and energy security have sparked a paradigm shift regarding the transportation of passengers and goods. Vehicle technologies are now changing radically towards higher energy efficiency and the use of alternative fuels. At the same time, transportation is increasingly managed in a smart manner, connecting a multitude of modes and making use of a variety of energy sources. Moreover, vehicles are becoming more and more assistive and autonomous, which will lead to even higher efficiency, fewer fatalities and more convenience. These developments are to be supported by novel services, processes and value creation, as well as by suitable public policies. LNMOB identifies the most promising conceptual work, groundbreaking findings and innovations from the industrial engineers and academic experts in this field. The overall aim of this series is the exchange of ideas to strengthen the scientific community, to create new value chains and to accelerate the move towards sustainable transportation systems (road, rail, water- and airborne) around the globe. Public research funding programs in Europe, America and Asia represent an important source of content. The selection of work for publication is done under the supervision of an international scientific advisory board that brings together some of the most influential technology leaders and scholars of our time. The target audience primarily consists of industry professionals and academic researchers working at the forefront of their fields, but the book series will also be of interest to advanced level and PhD students.

More information about this series at http://www.springer.com/series/11573

Gereon Meyer · Sven Beiker
Editors

# Road Vehicle Automation 8

Springer

*Editors*
Gereon Meyer
VDI/VDE Innovation + Technik GmbH
Berlin, Germany

Sven Beiker
Stanford University
Palo Alto, CA, USA

ISSN 2196-5544 ISSN 2196-5552 (electronic)
Lecture Notes in Mobility
ISBN 978-3-030-79818-5 ISBN 978-3-030-80063-5 (eBook)
https://doi.org/10.1007/978-3-030-80063-5



This Springer imprint is published by the registered company Springer Nature Switzerland AG
The registered company address is: Gewerbestrasse 11, 6330 Cham, Switzerland

# Preface

Without any doubt was 2020 an exceptional year, and 2021 continues to do so. Much has been discussed during those times in terms of what automated vehicles (AV) can contribute during critical situations like a pandemic and beyond. It was supposed that driverless vehicles would be very well suited to arrange for contactless delivery of goods and transportation of people. But, one also needs to attest that at a time where the "normal" traffic ceased, people kept distance all-around, and home delivery became the default; AVs were seemingly not yet ready to step up and fulfill their promise. This raises the question of how mature the promised "future of transportation" actually is. Rightfully so, as further progress in the technologies for environment perception, intention recognition and decision making, more real-world trials and further discourse on impacts and framework conditions will be needed to finally enter the intriguing world of level 4 automation.

In that sense, we are thankful for the contributions to this eighth edition of the Road Vehicle Automation books from our very much appreciated authors who tackle topics in technology, policy, business and other disciplines that all need to work together to make automated vehicles a reality. All those individuals make sure that we keep documenting the progress in road vehicle automation as it was discussed at the Automated Vehicle Symposium (AVS), held as a virtual event in July 2020. And while those contributions show great progress in their respective fields, we are also very appreciative of the unanswered questions and recommendations for further work that the authors point out.

While this book might feel a bit lighter than the previous seven volumes, we are heavily indebted to the contributors who dedicated their time during the unprecedented year that 2020 was. We do not want to rule out that, despite of the challenges of the pandemic, the consolidation that seems to have reached the automated vehicle sector might have an influence on the contents covered in this book; however, we also maintain that companies are moving closer to deployment and therefore relevant subjects get more proprietary and in return less talked about. To this effect, we absolutely observe that the field of road vehicle automation is as vibrant as ever, has more applicability than ever and might be as close to deployment as never before.

We are now looking forward to connecting with everyone at the 2021 Symposium and to keep documenting how we all get to "the future of transportation" in yet another volume of the Road Vehicle Automation books. In our personal views, this series is a great example of the international collaboration that is needed more than ever to find solutions for global challenges.

We are grateful to all authors for the time and efforts they spent on their chapters, to Jane Lappin, Valerie Shuman and Steven Shladover from TRB for their continuous support, and to colleagues at VDI/VDE-IT, particularly Laura Soto and Jacques Dalhoff, for assistance in the editorial process. Moreover, we would like to express our gratitude to the Springer teams in Heidelberg and Chennai for making possible this book publication despite of the challenges of the pandemic.

May 2021

Gereon Meyer
Sven Beiker

# Contents

# Introduction: The Automated Vehicles Symposium 2020

Steven E. Shladover[1](✉), Jane Lappin[2], and Valerie Shuman[3]

[1] University of California PATH Program, 1357 South 46th Street, Building 452, Richmond, CA 94804, USA
steve@path.berkeley.edu
[2] TRB Vehicle-Highway Automation Committee, Belmont, MA, USA
[3] Shuman Consulting Group, LLC, Skokie, IL, USA
vs@shumangroupllc.com

**Abstract.** The 2020 Automated Vehicles Symposium represented a significant departure from its predecessors, since it had to be converted on short notice from an in-person meeting to a virtual meeting in response to the global COVID-19 pandemic. Most of the originally planned content was retained in the process, although the activities were of necessity less interactive than in previous years. The plenary and poster presentations and breakout discussions continued to provide the meeting participants with the most up-to-date and authoritative information about the current international state of development and deployment of road vehicle automation systems, retaining its standing as the essential meeting for industry, government and research practitioners in the field.

**Keywords:** Road vehicle automation · Road transport automation · Automated vehicles · Autonomous vehicles · Self-driving vehicles

## 1 Overview

The 2020 Automated Vehicles Symposium was organized and produced through a partnership between the National Academies of Science, Engineering and Medicine (NASEM) Transportation Research Board (TRB) and the Association for Unmanned Vehicle Systems International (AUVSI), continuing the pattern established in the six preceding years. This meeting was organized to serve their constituencies' interests in understanding the impacts, benefits, challenges and risks associated with increasingly automated road vehicles and the environments in which they operate. It brought together key government, industry and academic experts from around the world with the goal of identifying opportunities and challenges and advancing Automated Driving System (ADS) research across a range of disciplines.

The symposium took place online over four days, 27–30 July, 2020, with session times scheduled to accommodate participants in Europe and east Asia as well as North America. Many speakers pre-recorded their presentations so that the presentation times were not constrained by their home time zones, and some of them participated in


G. Meyer and S. Beiker (Eds.): AVS 2020, LNMOB, pp. 1–11, 2022.
https://doi.org/10.1007/978-3-030-80063-5_1

roundtable question-and-answer sessions at more convenient times during the meeting. The plenary sessions were scheduled from 8–10 am, 12 noon–1 pm and 3–4 pm Pacific Time (corresponding to the originally planned physical meeting venue in San Diego), and all were recorded so that they were available for later viewing by registrants who could not be available at the scheduled times. Some of the speakers were available in real time to answer audience questions that were relayed through the online chat function.

Breakout sessions were scheduled in two-hour blocks to accommodate limited online attention spans. The early morning breakouts (5–7 am PDT) were convenient for international participants even if they were challenging for local participants. The other breakout time slots were at 10 am to 12 noon and 1 to 3 pm PDT, which meant that the east Asian participants had only one really awkward time slot (10 am to 12 noon) and everybody else was able to participate without too much difficulty.

The poster presentations were most adversely affected by the virtualization of the meeting, since it was more difficult to provide a suitably interactive experience with the online platform. The 55 posters were available for viewing throughout the meeting, and the poster presenters were available for online chat interactions with meeting attendees at several specific times.

The breakout sessions were each organized by committees of volunteers to address a wide range of topics. These were clustered into four thematic tracks to make it easier for attendees to identify the sessions of strongest interest to them:

- Policy and Planning
- Users and Human Factors
- Operations and Applications
- Technology.

In keeping with TRB practice, the plenary and breakout sessions were planned and produced by volunteers whose expertise and work informed the content of the sessions. In keeping with AUVSI practice, the production of the symposium was professionally managed by dedicated conference and logistics managers. The AVS20 Executive Committee reflected this mix of the two organizations:

Richard Bishop, Bishop Consulting; Robert Brown, TuSimple, Richard Cunard, Engineer of Traffic and Operations, TRB; Kevin Dopart, U.S. DOT Intelligent Transportation Systems Joint Program Office, Keely Griffith, Director of Industry Education, AUVSI, Jane Lappin, TRB Vehicle-Highway Automation Committee Chair; Steven Shladover, University of California PATH Program (and former chair of the TRB Vehicle-Highway Automation Committee); Valerie Shuman, Shuman Consulting Group, LLC and Chair, TRB CORVA Subcommittee, Edward Straub SAE, and Elizabeth Wilson, Perkins Coie LLP.

## 2 Keynote Talks

The symposium began with a keynote session featuring three safety experts from Waymo (Tracy Murrell, Interim Head of Safety, Qi Hommes, Head of System Safety, and

Matthew Schwall, Head of Field Safety) in a discussion moderated by Chris Gerdes from Stanford. They noted the importance of collaborations within the industry and between industry and government in order to develop standards, performance benchmarks and regulations that can protect both public safety and innovation. Success in deploying ADS depends not only on the technology to mitigate risks, but also on outreach to the public to earn public confidence. They cited their experience in developing robust internal processes for reviewing each software update for safety, providing the basis for advancing to the stage that they could confidently begin carrying some passengers in vehicles without onboard drivers in Chandler, Arizona (about 5 to 10% of the 1000 to 2000 rides per week that they were serving pre-pandemic). Waymo is participating in the development of industry standards that can support definition of safety assurance processes. Qi Hommes concluded the discussion by observing that automated driving is the most complicated technology to be developed to date.

Additional short keynote talks were provided on subsequent days of the symposium by Michael Kratsios, the Chief Technology Officer of the U.S. Government and James Owens, the NHTSA Deputy Administrator. Mr. Owens highlighted NHTSA's AV TEST initiative to encourage information sharing among the organizations that are testing ADS and their approvals of exemptions from current safety standards for a variety of automated vehicles in testing.

## 3 Plenary Panel Sessions

AVS20 continued the trend from AVS19 of devoting more of the plenary program time than in previous AVS meetings to panel discussion sessions on important topics featuring groups of speakers responding to questions from the moderator and interacting with each other, to break up the sequence of formal presentations. These also provided opportunities for audience members to submit questions through the online chat function.

### 3.1 Driver Monitoring and Management with L2 Automation

Level 2 driving automation systems require the driver to remain alert and ready to take back control of the vehicle. However, crash data and incident video show human drivers lose focus when L2 features are engaged, and dangerously over trust their capabilities. To better understand and address this safety threat, Jane Lappin moderated a panel of experts in automotive safety systems and human-machine interaction: Alexandra Mueller, Insurance Institute for Highway Safety, Brad Stertz, Audi of America, Bryan Reimer, MIT Center of Transportation and Logistics, and Ensar Becic, Office of Highway Safety in the National Transportation Safety Board. The panelists spoke from their research with currently available L2 systems to describe the safety problem and its consequences. Drawing from naturalistic driving studies and crash investigations, they agreed that future L2 and L3 systems must incorporate driver management methods that encourage sustained driver engagement with vehicle operations and limit the operating environments where these systems may be used.

### 3.2 UL4600 Safety Cases: Industry Approach and Applications

Junko Yoshida from AspenCore Media moderated a discussion among representatives of Edge Case Research, who led the development of the UL4600 safety case standard and the first industry users of the standard from Uber ATG and Liberty Mutual Insurance. Phil Koopman provided an overview of how UL4600 provides a framework for defining a system-level safety case, complementing other standards that address individual safety elements. Chris Mullen explained how Uber ATG used it to update their company's safety case and Ben Lewis explained how Liberty Mutual uses it to assess the safety cases of AV developers so that they can set insurance rates based on a realistic estimate of risks.

### 3.3 Lessons Learned from Uber Crash

Kristin Kingsley, KKingsley Consulting, moderated a discussion with Ensar Becic, National Transportation Safety Board, and Christopher SanGiovanni, Director of Organizational Safety Management, Uber ATG, that reviewed the lessons learned from NTSB investigation of the 2018 fatal Uber crash in Tempe, Arizona, and the safety protocols that Uber ATG have since developed and now enforce. Drawing from the NTSB investigation, Dr. Becic recommended that state regulators and NHTSA implement mandatory safety management systems, review, and approvals for on-road ADS testing to better mitigate safety risks associated with crashes and operator inattentiveness, and to address the appropriateness of company countermeasures for testing conditions. Mr. SanGiovanni reviewed the comprehensive internal and external reviews undertaken by Uber ATG following the crash, stressing the foundational importance of company safety culture. He described the Uber ATG Safety Management System that was developed to increase operating safety in all domains and specifically for on-road testing. Mr. SanGiovanni cited the importance of the SAE Automated Vehicles Safety Consortium as a forum for implementing lessons learned towards the development of safety standards.

### 3.4 Investment and Capital Updates

Venture capitalists play a critical role in the trajectory of AV development, providing money and wisdom to guide start-up companies to maturity. Gretchen Effgen, VP of Go-to-Market and Marketing at Hyundai-Aptiv Autonomous Driving Joint Venture, facilitated a lively discussion about the current state of the AV industry with Eran Sandhaus, Managing Director, Copia Growth Partners, Jim Scheiman, Founding Managing Partner, Maven Ventures, and Orin Hoffman, Venture Partner, The Engine. The discussion ranged through industry consolidation, ROI timelines, pandemic food delivery, the role of infrastructure in reducing the complexity of the operating domain, the role of the regulatory framework, and areas of the AV industry that are still ripe for investment.

### 3.5 Why Truck Fleets Will Lead the Way for AV Deployment

Selika Talbott from American University moderated the discussion with representatives of AV truck developers Locomation, TuSimple and Plus.ai and potential customer UPS.

The panelists expressed general agreement that the labor impacts of truck automation would be minor in scale and gradual in evolution, and all were amenable to strong federal regulations on AV safety to ensure that the entire industry would be safe enough to earn public confidence. UPS is most interested in use of lower levels of automation to improve safety, while the AV system developers are interested in the combination of safety and efficiency gains from higher levels of automation.

### 3.6 Current Status of Federal AV Legislation

Hilary Cain, Alliance for Automotive Innovation, led a conversation with three legislative affairs colleagues exploring the issues that have delayed passage of federal AV legislation: Jamie Boone, Consumer Technology Association, Peter Kurdock, Advocates for Highway & Auto Safety, and Ron Thaniel, Intelligent Transportation Society of America. They agreed on the need for a comprehensive bill that can support innovation and competition, preserves federal authority over the vehicle and state and local authority over the road, and puts in place a national framework for safety plans for AVs. What are the top legislative issues that remain unresolved? Liability, arbitration, the number of vehicle exemptions, state preemption, workforce displacement, additional funding to NHTSA, cybersecurity, and access to data for NTSB that can be used to assess safety operations. They closed in agreement that the pandemic and the elections would further delay legislative action on automated vehicles.

### 3.7 Hard Truths from Journalism's Best

Grayson Brulte, Brulte & Company, led a lively discussion exploring the state of the industry with Alan Ohnsman, Forbes, Joann Muller, Axios Navigate, and Kirsten Korosec, TechCrunch. Mr. Brulte opened the conversation asking, "Where are we?" The journalists replied, not as close to L4/5 as the industry predicted. There's still a lot of enthusiasm among funders for the technology, but deals are available only for companies with a clear business strategy and strong management. The pandemic has accelerated the consolidation trends. The absence of coherent regulation has cost lives. What about AV truck start-ups? Like the other independents, they've got to have a major tech partnership or a major OEM partner to make it. Will Amazon treat Zoox like Zappos, allowing it to remain independent? Too soon to tell. Amazon is building a massive logistics company that may be just for their own use, or it may be a future stand-alone service. How will the interests of companies like Walmart affect development? Keep an eye on them for opportunities to leverage AVs for shipping, and maybe also for customer shuttles. In closing, they agreed that there would be at least several major deals in the offing for the remainder of the year.

### 3.8 Remote Support to Accelerate ADS Deployment

Richard Bishop facilitated a wide-ranging discussion that explored ADS teleoperation with Amit Rosenzweig, Ottopia, Elliot Katz, Phantom Auto, Manuela Papadopol, Designated Driver, and Pär Degerman, Einride. Current developments in remote assistance and teleoperations are increasing the distance between the automated vehicle and human supervisors, and increasing the ratio of vehicles to driver, creating new safety challenges and new business opportunities. The panel explored the conditions for trade-off in control between the ADS and the remote driver, and discussed the requirements for cybersecurity, safety, and the need to develop standards and regulations. They identified workforce impacts and consumer acceptance as key societal issues. They agreed that 5G is important for remote driving but less critical for remote assistance where waypoints can be used to orient the ADS. The right ratio of tele-operators to vehicles depends on the maturity of the ADS and the complexity of the operating environment. Pär Degerman, Einride, summarized by saying, "We'll always have humans in the loop."

## 4 Plenary Presentations

The individual plenary presentations were distributed across the plenary program in combination with the panel discussions to avoid Zoom fatigue from too long a sequence of consecutive presentations. These were all pre-recorded to ensure reliable timing and to avoid surprises from connectivity problems, and remained available for viewing at any time after their initial scheduled time. The presentations are grouped here in broad thematic categories:

### 4.1 Research and Technology Topics

- Bryan Reimer, MIT – Observations from the MIT AVT Naturalistic Driving Study
- Sven Beiker, Silicon Valley Mobility – Sensors for Automated Vehicles – Recent Technology and Market Trends
- Lutz Eckstein, RWTH Aachen – Pegasus VVM: Verification and Validation Methods for Level 4 Automated Driving

### 4.2 Automated Driving Applications

- Lindsay Wiginton, City of Toronto – Transit-centered Automation in Canada's Largest City: Toronto's AV Future
- Jordana Maisel, IDEA Center – Ensuring Community Mobility for All
- Emily Weslosky, Nuro – The Path to Contactless Delivery and AV Operations at Scale
- Nadeem Sheikh, Lyft – Greater Mobility in a Post-COVID World

### 4.3 National and Multi-national Programs of Automated Driving R&D and Regulation

- Finch Fulton, U.S. DOT – U.S. DOT Automated Vehicle Research Activities
- Tom Alkim, European Commission – Connected, Cooperative Automated Mobility (CCAM), the European Approach
- Iain Forbes, UK Department for Transport – Automated Vehicles in the UK
- Marcus Burke, NTC Australia – How Do We Regulate Automated Vehicles Out on the Road?
- Seigo Kuzumaki, SIP-adus Program –The Challenges for Automated Driving Systems Realization in Japan
- Niels de Boer, CETRAN – Singapore AV Trial Testbed

## 5 Breakout Sessions

AVS Breakouts gather key experts from around the globe for more in-depth consideration of specific topic areas. As smaller gatherings, they are more interactive than the plenary sessions, providing ample opportunities for conversation and collaboration.

The goal of the breakout sessions is to answer the questions: What needs to be true to make the AV vision become a reality? How can our research help drive progress year on year? The 2020 sessions covered a wide range of specialized topics from across the field to enable this discussion for the industry as a whole (see program list below).

The primary findings from each afternoon's breakout discussions were reported back to the plenary the following morning. The combined summaries provided in these Daily Roundups distill the latest insights from across the industry, including:

- Collaboration is fundamental. This was the single most consistent message from the 2020 breakout discussions: development and deployment of AVs must be a truly cross-cutting effort. Nearly every breakout group called for increased levels of partnership, be it cross-domain, cross-sector, cross-industry, cross-jurisdiction and/or global. Similarly, there was a consistent call for standards of every kind, from policy to technology.
- We have a major system-level optimization opportunity. There is growing energy behind the idea that AVs have the potential to benefit not just individual consumer safety, but the transportation network and society as a whole.
- We need to focus on integration & trust. AVs require a significant "leap of faith" on the part of the humans that interact with them, and it is critical that we as an industry effectively address this challenge both within single vehicles (at all levels of automation) and across the entire network. AVs must be good citizens or they will not be accepted.
- This is a multi-decade marathon. In contrast to earlier aggressive predictions of near-immediate mass deployment, experts now recognize that achieving an AV future will take many years. It is increasingly important to find ways to focus on the small steps that move us forward, in a coordinated way.
- There is a lot to learn from the pandemic. In particular, there is now a renewed focus on flexible and equitable AV use cases; automated trucking and delivery that supports supply chains under stress and contactless consumer interaction; and solutions that we can deploy now, specifically L2 automated vehicles.

### 5.1 2020 AVS Breakout Sessions

#### 5.1.1 Policy and Planning Sessions

- Regulatory Policy for Automated Vehicles
- No-Regret Options for Policy Making and Infrastructure Development
- Planning for Automated Vehicles: How to Plan for an Unknown Future
- Local Roadmaps for Autonomous Vehicles: A Workshop to Develop Local Strategies
- Mock Trial: AV Cyberattack: Who Pays for the Damages?
- Energy & Environmental Implications of Connected & Automated Vehicles
- The (Dedicated) Road to Deployment: What are the Priorities?

### 5.1.2 Users and Human Factors Sessions

- Driver Monitoring in Future AVs: Identifying and Mitigating Fatigue and Distraction in L2/L3 Driving
- Human Factors in Regulatory Frameworks
- The Potential for AVs to Support Active Aging and Community Mobility in Suburban and Ex-urban Areas
- Mental Models of Automated Driving

### 5.1.3 Operations and Applications Sessions

- Using Connected Devices to Improve Work Zone Data for Work Zone Data Exchanges
- State of the UX Research: ADA30 and the Complete Trip Coalition
- Autonomous Shuttles and Buses: From Demonstrations to Deployment
- Delivering Freight on Automated Trucks Today
- Roadway Capacity Effects of CAV's: How Much and When?
- Garbage In, Garbage Out: Building Robust Models of CAV Performance in Traffic
- Real World AV Data – Collection to Sharing and Everything In Between
- Integrating Emergency Response into Connected Infrastructure Systems and Automation
- Designed, Wheeled, Delivered: Emerging Concepts for Automated Urban Delivery Vehicles
- The State of Open Data and Training Data to Advance Automated Vehicle Research
- Reading the Road Ahead: Global Efforts Toward Assessing Infrastructure's Ability to Support ADAS and HAV Operation

### 5.1.4 Technology Sessions

- Safety Assurance of Automated Driving
- Security and Privacy of Automated Driving: Hot Topics
- Public and Private Sector Collaboration to Advance Automated Driving Systems Testing and Deployment
- Enabling Technologies - A Peek Under the Hood
- AI for AV Control and Traffic Operations: Challenges and Opportunities
- Shark Tank – Open Debate
- Advancing the Intelligent Transportation Systems (ITS) Reference Architecture with Information Views
- On-Road Automated Driving Standards Priorities and Emerging Work Topics
- New Simulation Tools for Training and Testing Automated Vehicles
- Collaborating with CARMA: Case Studies from FHWA's CARMA University Partners & Workshop on Accelerating Research with CARMA

## 6 General Cross-Cutting Observations

As the field of road vehicle automation has advanced and the level of knowledge of the issues has grown over the past several years, the areas of emphasis within the Automated

Vehicles Symposium have continued to evolve. In this most recent meeting, several general observations are worth noting:

- Many speakers emphasized the importance of industry collaborations and cooperation to be able to move ADS technology forward. There are so many technical and organizational challenges, requiring such a broad multi-disciplinary mix of skills, that even large companies cannot do this on their own. The companies that are not yet aligned with partners will need to get their partnerships lined up within the coming year in order to remain viable.
- Multiple speakers noted the need for ADS technology developers and vehicle manufacturers to work together very closely on the implementation of the technology in the vehicles. This kind of tight integration cannot be done with retrofits of ADS into vehicles, so the retrofit concept appears to have faded from serious consideration.
- There is also a need for strong cooperation among industry, government (at national and regional/state/local levels) and the general public (through public interest groups) in order to make progress toward realistic and intelligent regulatory frameworks and public acceptance of the safety cases for the automation systems.
- The U.S. political situation is not currently conducive to reaching agreement on a national framework for regulation of ADS, but many speakers also noted that development of industry consensus standards and government regulations are important milestones toward reducing uncertainty for industry developers of ADS. Smaller countries with a higher level of trust between government and industry will be better positioned to overcome this.
- Speakers generally agreed that the existing regulatory frameworks are not a good fit for ADS, but at the same time they recognized that it is difficult and time consuming to create new regulatory frameworks. This is especially challenging when there are large differences of opinion between industry and public interest groups, as there are in the U.S., regarding the degree of regulation that will be needed on ADS safety.
  - The developers of ADS for trucking were more amenable to having strong government regulations on ADS safety to ensure that their entire industry is safe enough to earn public confidence than the ADS passenger vehicle developers have been.
- The huge technological complexity of ADS was more openly acknowledged by major industry participants than in the past, when there was more of a tendency to downplay the technological challenges.
- Speakers talked about how large the leap is from the current generation of Level 2 driving assistance systems to driverless Level 4 systems. More companies have come to recognize this as they have prepared to make that leap themselves, and have noted that this leap needs to begin within very narrowly defined ODDs in order to limit the complexity of the challenge.
- The current Level 2 automation systems can be viewed as "dress rehearsals" for higher levels of automation, revealing the issues of technology performance limitations and ineffective user interfaces that will have to be improved significantly for more highly automated systems.
- Significantly more work needs to be done on interactions with the general public, who will be the end users of the technology, to understand their needs and concerns on the one hand and to educate them about the capabilities and limitations of the

ADS technologies on the other hand. The need for education (and for more effective human interface designs) has been brought into focus by the problems with Level 2 automation systems that are not properly understood by consumers, leading to safety problems.

- There was a strong trend toward recognizing that for the higher levels of automation, the goods movement applications are likely to precede the people movement applications for a variety of reasons associated with business cases, technological complexity and safety. This trend has been accelerated by COVID-19, with the increased demand for contactless package delivery and the increased anxiety about people sharing rides. There was a lot more enthusiasm about package delivery robots than about automated ride hailing services ( "robotaxis") as a result.
- The "robotaxi" applications were increasingly recognized to be very challenging to implement, after huge investments over multiple years by several major developers have produced very limited actual service on the road (a limited number of Waymo rides in Arizona). Investors may be losing patience about the length of time needed to achieve a return on investment.
- Although there have been many small-scale trials of low-speed driverless passenger shuttles, these have all required the use of onboard attendants as well as remote assistants. The economic model is not viable for general use under these conditions, yet none of the people who spoke about these shuttle operations could realistically consider eliminating the onboard attendants.
- Concerns about workforce displacement associated with elimination of driving jobs were significantly lower than in prior years, based on growing recognition of how gradual the rollout of driverless ADS is likely to be.
- There was a new willingness to talk about the vital role of remote human monitoring and assistance for driverless ADS. Virtually all the organizations developing and applying driverless ADS agreed that remote monitoring and support would be essential, but only a few showed interest in remote driving (and that only for low-speed operations). The remote monitoring and assistance jobs were noted as excellent employment opportunities for former drivers, with more attractive working conditions.
- There was considerable discussion about how the COVID-19 pandemic has affected progress on ADS, with general agreement that it was accelerating several trends that were already evident:
  - Consolidation of companies, continuing the shake-out of those with weaker financing, technology or business models and forcing companies to form strategic partnerships
  - Shifting emphasis from people movement to goods movement
  - Reduced interest in shared people movement in particular, raising concerns about whether ADS might exacerbate the trend toward low (to zero) vehicle occupancy
  - Raising the already-high uncertainty level surrounding most aspects of automated driving (a concern for decision makers in all stakeholder categories associated with ADS)

- As in other recent years, the most important unresolved issues were associated with ADS safety, and this year the issues of concern began to come into stronger focus:
  - Recognizing that the safety considerations are not only technical, but also involve more deep-seated issues of organizational culture for the organizations that are developing and deploying ADS.
  - Recognizing that everybody needs to consider both the actual risks and the perceived risks because both affect decision making by relevant stakeholders.
  - Decisions need to be made at the societal level about the acceptable level of ADS safety and about the baseline point of reference for comparison.
  - Many different approaches have been suggested for safety verification, but it has not yet become clear how to combine them into an integrated approach that takes best advantage of the strengths of each approach.
  - The technical challenges to safety verification remain daunting and will require extensive and coordinated development efforts internationally because of the complexity of the software and sensor technologies, the cost and time that will be needed to implement safety verification, and the difficulty of developing and validating sufficiently realistic simulations to provide a cost-effective platform for safety verification.

# Part I: Public Sector Activities

# The Challenges for Automated Driving Systems Realization in Japan; SIP-adus

Seigo Kuzumaki(✉)

Toyota Motor Corporation, 1, Toyota-cho, Toyota, Aichi 471-8572, Japan
seigo_kuzumaki@mail.toyota.co.jp

**Abstract.** SIP, or Cross-ministerial Strategic Innovation Promotion Program is a 5 year R&D program led by the Japan government. SIP-adus, or automated driving system for universal service is a project for the realization for automated driving systems (ADS) with Government-Industry-Academia cooperation. We tackle various issues such as R&D, regulation reform, public acceptance creation, international cooperation for realization of cooperative ADS together. The project focused on cooperative R&D themes such as Dynamic Map, Safety assurance, Cybersecurity, and so on. The status of several examples of our challenge will be reported.

**Keywords:** Automated driving · Automated vehicles · Dynamic Map · Impact assessment · Field Operational Test · Cyber security

## 1 Society 5.0

Japan has its particular challenges for digital transportation of manufacturing.

Society 5.0 was proposed in the 5th Science and Technology Basic Plan as a future society that Japan should aspire towards. By a high degree of convergence between cyber space and physical space, Society 5.0 will be possible to achieve a society that can both promote economic development and find solutions to social problems.

In Society 5.0, a huge amount of information from sensors in physical space is accumulated in cyberspace, and then this big data is analyzed by artificial intelligence (AI) (Fig. 1).

Automated driving system (ADS) is regarded as one of core technologies in Society 5.0. In SIP-adus, we are trying to establish a traffic environmental database for ADS, and to promote data collaboration among industry and government.

New values are expected to be generated through AI analysis of big data in a database spanning diverse types of information including sensor data from automobiles, real-time information on the weather, traffic, accommodations, and personal history.


G. Meyer and S. Beiker (Eds.): AVS 2020, LNMOB, pp. 15–21, 2022.
https://doi.org/10.1007/978-3-030-80063-5_2

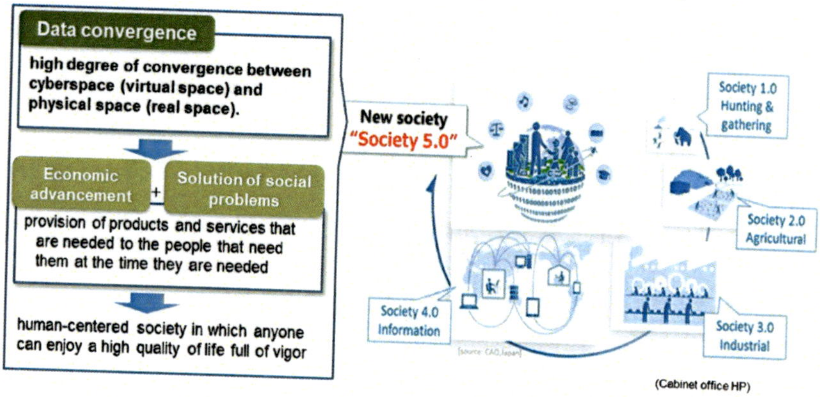


**Fig. 1.** Society 5.0

## 2 Dynamic Map

ADS needs to estimate its own position accurately, which called 'localization', and find out and follow a safe path forward. To achieve this, ADS uses high definition 3D digital map and on-board sensors like camera, radar and so on. The vehicle estimates its own position accurately by comparing with sensor data and HD digital map data which is installed onboard vehicles (Fig. 2).

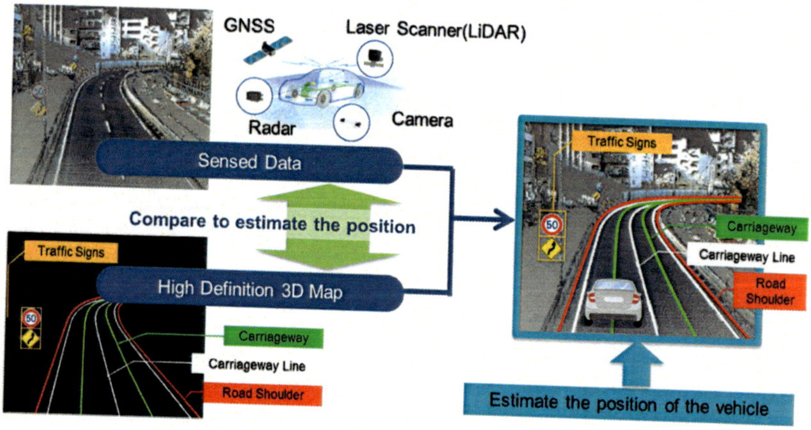


**Fig. 2.** Vehicle position detection using HD 3D map

So HD digital map is quite important for ADS. And also other information like traffic rules, and so on. Figure 3 shows a concept of Dynamic Map. Dynamic Map consists of high definition 3D map and dynamic data. It is conceptually composed of four layers: static data, semi-static data, semi-dynamic data and dynamic data.

This Dynamic Map database is thought to be useful not only for ADS, but also for all other vehicles and drivers on the road. 3D map data should be fresh, the easiness for updating is important. Also, scalability and low cost are required.

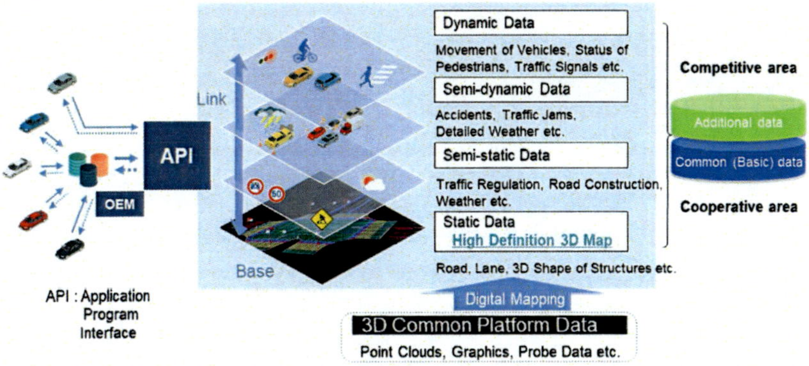


**Fig. 3.** Dynamic Map

As the result of first phase of SIP-adus activity, Dynamic Map Platform Co. Ltd (DMP) was established in 2017. Six map companies and nine automakers invested in DMP has released HD 3D map data of all exclusive motor way in Japan on commercial basis from 2018.

In second phase of SIP, we are developing traffic environmental data, such as traffic signal information, merging area traffic information, lane level traffic jam information created by using vehicle provided data. We think this will lead to realize the cooperative ADS and finally Society 5.0, and new data business creation.

## 3 Field Operational Test in Tokyo Water Front Area

We have started FOT (Field Operational Test) in the Tokyo waterfront area from November 2019. In the Tokyo waterfront city area, we provide signal information from about 30 traffic lights by DSRC. In Haneda area, we are conducting FOT for the next generation Advanced Rapid Transit. On the metropolitan express way, we provide traffic information on the main lane for merging assistance.

The main purpose of the FOT is the validation of ADS, traffic environmental data and so on, under real environments on public roads. Another purpose is to enhance international cooperation and harmonization. We have 29 participants in our FOT, which include car manufacturers, suppliers, universities, venture companies not only from Japan but also from Europe.

We are now getting good feedback from participants regarding effectiveness of signal information to ADS, format of the data and so on. COVID-19 affected our FOT schedule, but we will complete this fiscal year schedule testing by March 2021, and soon after, its interim report will be released. We got many requests for extension from participants, so we decided to keep this FOT condition by the end of March 2022 with added new traffic environmental information (Fig. 4).

**Fig. 4.** Field Operational Test in Tokyo waterfront city area

## 4 Safety Assurance

Safety is critical and essential for ADS realization. In order to realize and spread ADS in the market, it is necessary for us to have a feasible methodology to assess the safety performance of the vehicle.

SIP is working to develop a platform to evaluate the safety of ADS in a virtual space. The project is named DIVP™, Driving Intelligence Validation Platform. DIVP™ consortium is constructed from 8 companies and 2 universities (Fig. 5).

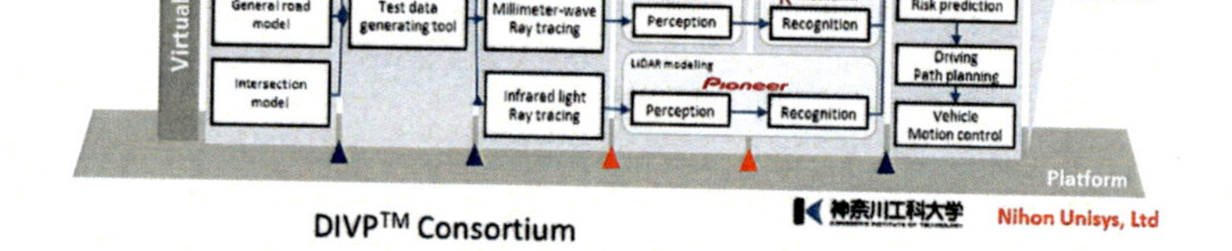


**Fig. 5.** The structure of DIVP™

When driving, the human driver is doing "recognition", "judgement" and "operation" continuously. ADS have to do these by itself on behalf of the human driver. Thus, sensor performance is critical for ADS. This project focuses on a precise duplication from real to virtual, and on sensor models' verification of consistency with real word testing. DIVP™

scope covers especially, "Physical Model improvement" & "Computing Performance" in Trinitarian approach (Fig. 6).

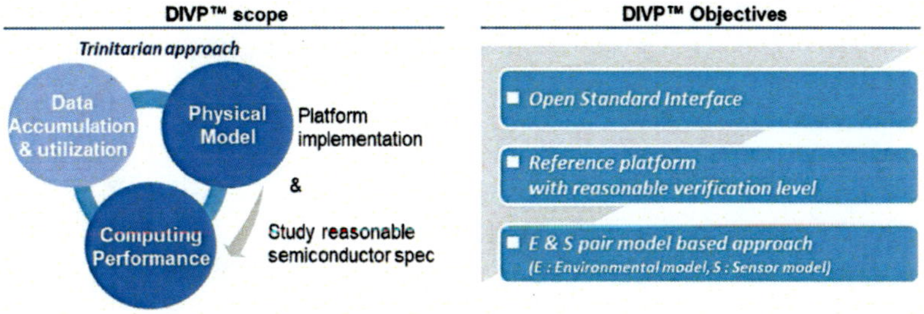


**Fig. 6.** Scope and objective of DIVP™

Based on this approach, DIVP™ objectives are to define open standard Interfaces, to establish 'reference platform' with reasonable verification level, (especially, for sensor modeling), and to establish the Environment & Sensor pair model-based approach for Validation & Verification reality (Fig. 7).

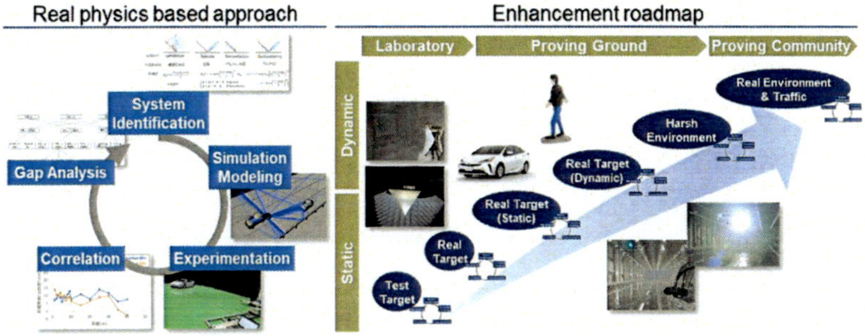


**Fig. 7.** Physical modeling framework DIVP™

We think that process, which are System Identification, Simulation Modeling, Experimentation, Correlation, and Gap Analysis is important for models' consistency verification. Basic verification is being conducted at the laboratory level and at the proving ground. This includes static and dynamic verification.

Finally, DIVP™ plan to carry out extended verification of the scenario by determining the region.

## 5 Cybersecurity

Cybersecurity is another important issue for ADS, which will have a connected function. In first phase of SIP, we tried to establish the penetration testing method for vehicle. Currently new methods and techniques for cyber-attacks are introduced continuously, all after-sales vehicles are under threat of the new cyber-attacks.

Intrusion detection system, or IDS, are becoming popular as a measure against the new cyber-attacks for after-sales product. There are many components and solutions of IDS by various vendors in the market, and generally it is difficult for the user (OEM) to find which IDS is best or enough. Therefore, in the second phase of SIP, we try to establish evaluation method for IDS components/solutions provided by various security vendors from the view point of the user (OEM) (Fig. 8).

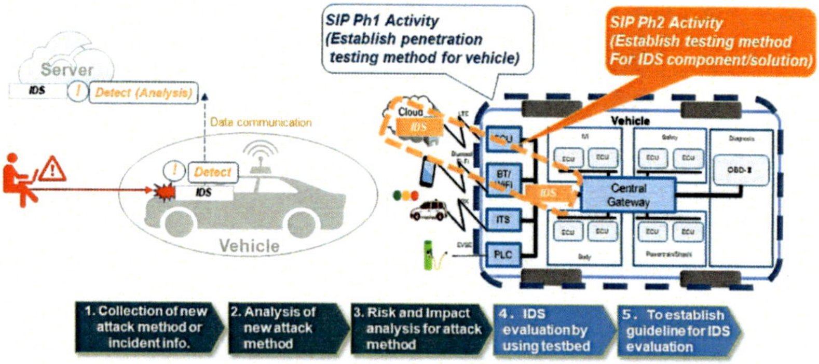


**Fig. 8.** Cover area of second phase of SIP-adus cybersecurity

There's a strong need in the automobile industry to link quickly to respond/recover after detection. Only "detection and alerts" are not enough.

One of the themes of the 2nd stage research for us is what the technical elements necessary for the initial action of respond/recover of IDPS (Intrusion Detection Protection System) are. Another one is formulation of guidelines for IDS evaluation. Our 2nd stage research just started and it will continue until the Mid of 2022.

## 6 Conclusion

The Second phase of SIP, which started since 2018, has passed the halfway point. We tackled to solve the cooperated issues regarding R&D, regulation and public acceptance for ADS realization with industry-government-academia cooperation. SIP is a program which aims to promote intensive R&D from fundamental research to practical and commercialization, so we promote our R&D with milestones in order to produce output within the last 2 years.

But, from a practical point of view, it needs a long time to realize high level of ADS under various operational design domain. So, we think that it also need to leave a legacy for ADS R&D framework with industry-government-academia cooperation. It's our challenge for the future.

## References

1. Sugimoto, Y., Kuzumaki, S.: SIP-adus: an update on Japanese initiatives for automated driving. In: Meyer, G., Beiker, S. (eds.) Road Vehicle Automation 5. Springer, Cham (2019). https://doi.org/10.1007/978-3-319-94896-6_2

2. Minakata, M.: SIP-adus filed operational test. In: SIP-adus Workshop (2020)
3. Inoue, H.: Driving intelligence validation platform. In: SIP-adus Workshop (2020)
4. Uehara, S.: Cybersecurity. In: SIP-adus Workshop (2020)
5. Okuyama, K.: Research for effectiveness and technology of intrusion detection systems. In: SIP-adus Workshop, Tokyo (2020)

# Policy and Regulation of Automated Vehicles: Spotlight on U.S. Federal and States

Mollie D'Agostino[1(✉)], Kelly Fleming[1], Kristin White[2], Marc Scribner[3], and Baruch Feigenbaum[3]

[1] University of California Davis, Davis, USA
{mdagostino,kelfleming}@ucdavis.edu
[2] Minnesota Department of Transportation, Saint Paul, USA
kristin.white@state.mn.us
[3] Reason Foundation, Washington, DC, USA
{marc.scribner,baruch.feigenbaum}@reason.org

**Abstract.** This chapter outlines the key developments and issues in the U.S. national, state and local automated vehicle (AV) policy landscape. There are a range of perspectives among U.S. government and private-sector stakeholders. Despite the wide range of viewpoints, the authors aim to underscore where there is agreement on key policy issues, and where there are opportunities for more discussion. Federal policy to date has focused on a voluntary system for ensuring safety of AVs. Federal regulatory actions have been limited to guidance documents, although the government recently proposed a safety framework for the automated driving system (ADS). States have varied in their approaches to AV policy and this chapter summarizes research coding state activities on a spectrum from more restrictive, such as California, to more permissive, such as Arizona. In the final section of the chapter the authors discuss the role of interstate coalitions. This Chapter is organized into five sections:

- Section 1 Introduction
- Section 2 National Landscape for Federal AV Legislation and Regulation
- Section 3 State Automated Vehicles Policy and Regulation
- Section 4 Regional Coalitions
- Section 5 Conclusion

## 1 Introduction

Automated vehicles (AVs) could bring significant societal benefits if they succeed in improving safety outcomes over driver-operated vehicles. AVs could bring new mobility options to the elderly, disabled, and underserved individuals improving job accessibility and event health outcomes. AVs could expand access to jobs, improve economic efficiency, and reduce traffic congestion. But these benefits cannot be realized without industry innovation and governance policies. A balanced and predictable policy landscape will enable these AV best case scenarios outcomes, and guardrails may be


G. Meyer and S. Beiker (Eds.): AVS 2020, LNMOB, pp. 22–39, 2022.
https://doi.org/10.1007/978-3-030-80063-5_3

necessary to avoid a future in which AVs worsen transportation problems and contribute to worsening congestion, and additional safety risks.

This paper describes the U.S. Federal and State Policy landscape for AVs. It builds on a session at the Automated Vehicle Symposium in July 2020, titled Regulatory Policy for Automated Vehicles. While we do not capture every topic discussed in this symposium breakout, we aim to unpack the main themes and present a balanced account of industry, government, and academic perspectives.

## 2 Landscape for Federal AV Legislation and Regulation

### 2.1 Summary of U.S. Department of Transportation (USDOT) Guidance and National Testing Regulatory Context

Federal regulators have not set binding regulations for automated vehicles. Instead, development and testing are occurring under a voluntary scheme shaped by a series of guidance documents. However, the future of this approach remains uncertain as regulators continue to work on these issues. The National Highway Traffic Safety Administration's (NHTSA) released a Framework for Automated Driving System Safety [1], an advance notice of proposed rulemaking (ANPRM) in November 2020. This AV Framework ANPRM, which was published in the public record in early December 2020, seeks to develop a safety framework for the automated driving systems (ADS) distinct from the Federal Motor Vehicle Safety Standards (FMVSS).

The FMVSS sets requirements for vehicles, whereas with this AV Framework ANPRM, NHTSA is proposing a potential safety framework that would apply to the software and hardware of the ADS. In doing so, NHTSA aims to address two key aspects of ADS safety: process measures and engineering measures [1, p. 9]. According to NHTSA,

> *The process measures (e.g., general practices for analyzing, classifying by severity level and frequency, and reducing potential sources of risks during the vehicle design process) would likely include robust safety assurance and functional safety programs. The engineering measures (e.g., performance metrics, thresholds, and test procedures) would seek to provide ways of demonstrating that ADS perform their sensing, perception, planning, and control (i.e., execution) of intended functions with a high level of proficiency* [1].

This proposed safety framework is a departure from the previous guidance documents issued by the agency, in that it is specific and actionable proposed AV regulation, created after more than seven years of extensive discussion. NHTSA developed a Preliminary Statement of Policy Concerning Automated Vehicles in 2013, but the first instance of the USDOT wading into AV policy guidance began in 2016, when NHTSA issued the Federal Automated Vehicles Policy (FAVP). A key element of this first guidance document was that NHTSA clarified four tools that are available for regulating AVs, 1) Letters of interpretation; 2) Exemptions from existing standards; 3) Rulemakings to amend existing standards or create new standards; and 4) Enforcement authority to address defects that pose an unreasonable risk to safety. The FAVP also contained guidance for states and a 15-point Safety Assessment [2].

In 2017, NHTSA released Automated Driving Systems 2.0 (ADS 2.0), which expanded interpretations from the previous document, and included guidance for state governments to take action. A key outcome of AV 2.0 was that it replaced the 15-point Safety Assessment in the FAVP establishing a new voluntary mechanism for assessing the safety of AVs. AV 2.0 introduced 12 safety elements and a set of Voluntary Safety Self-Assessments (VSSAs). Subsequent guidance documents built on ADS 2.0 [3].

In October 2018, the U.S. Department of Transportation (DOT) released a third guidance document called Automated Vehicles 3.0 (AV 3.0). Unlike the ADS 2.0 guidance released in 2017, AV 3.0 built upon, rather than replaced, existing automated vehicle guidance. Departing from the previous narrow focus on light-duty automated road vehicles, AV 3.0 provided DOT-wide guidance across all modal operating administrations [4]. An important contribution of AV 3.0 was that it announced future rulemakings designed to modernize terms such as "driver" and "operator" to be inclusive of non-human direction, such as through automated driving systems. As a resource, AV 3.0 also included an appendix containing potentially relevant technical standards and information related to ongoing automation standardization efforts, allowing interested readers to more easily observe and track the technical standards development progress [4].

AV 3.0 also announced a variety of administrative actions planned by DOT's modal agencies. These include:

- The Federal Motor Carrier Safety Administration's advance notice of proposed rulemaking (ANPRM) on regulatory modernization for commercial automated vehicles [5].
- The Federal Railroad Administration's research initiative on a concept of operations for the use of automated and connected vehicle technologies to improve safety at grade crossings [6].
- The Federal Transit Administration (FTA)'s research plan on automating bus transit [7]. Since this initial effort was launched the FTA has published several updated reports, the most recent of which was published in Sep 2020 [8].
- The Maritime Administration's investigation into the potential uses of automated vehicle technology to relieve heavy-truck congestion at ports [9].
- **The National Highway Traffic Safety Administration's notice of proposed rulemaking** on possible exceptions from Federal Motor Vehicle Safety Standards that presume a human driver is present and a request for comments on a streamlining and modernizing the FMVSS exemption process.
- Federal Highway Administration (FHWA) can pursue research this could include **technical assessments and traffic modeling related to safety, traffic management and energy impacts. FHWA can also incorporate findings into the Manual on Uniform Traffic Control Devices (MUTCD) and review the existing Uniform Vehicle Code** (UVC).

In January 2020, USDOT and the National Science and Technology Council released Automated Vehicles 4.0 (AV 4.0). Similar to AV 3.0, AV 4.0 augments and complements existing guidance rather than replacing it. Similar to AV 3.0, AV 4.0 expands the scope of automated vehicle guidance, in this case from AV 3.0's USDOT operating administrations to the entire Executive Branch. While not as technically focused as previous automated vehicle guidance documents, AV 4.0 provides ten policy principles to guide automated vehicle policy development across the federal government. Table 1 details these recommendations.

In January 2021, USDOT issued its Automated Vehicles Comprehensive Plan, which built on the principles of AV 4.0 and summarized federal policy development activities. January also saw NHTSA release a draft final rule on Occupant Protection for Vehicles with Automated Driving Systems. Under this rule, NHTSA would amend 13 of its 25 crashworthiness 200 Series FMVSS to address barriers to self-certifying ADS-equipped vehicles that NHTSA identified. However, the change in administrations in late January triggered a "regulatory freeze" of pending rulemaking actions across the federal government. NHTSA subsequently removed the draft rule and accompanying press release from its website. To date, the final rule has not been published and its future remains uncertain.

**Table 1.** Policy themes and corresponding principles from AV 4.0

| Policy theme | Policy principle |
|---|---|
| I. Protect users and communities | 1. Prioritize safety |
| | 2. Emphasize security and cybersecurity |
| | 3. Ensure privacy and data security |
| | 4. Enhance mobility and accessibility |
| II. Promote efficient markets | 5. Remain technology neutral |
| | 6. Protect American innovation and creativity |
| | 7. Modernize regulations |
| III. Facilitate coordinated efforts | 8. Promote consistent standards and policies |
| | 9. Ensure a consistent federal approach |
| | 10. Improve transportation system-level effects |

### 2.1.1 How Does the U.S. Regulatory Approach Compare to Select International Examples?

Different countries are examining AV policy and regulation with different priorities and goals. For example, according to a World Economic Forum report, the UK is considering a *scenario library* that could establish safety across operators, although they currently require test drivers in automated vehicles [10].

Singapore developed a tiered approach with three *milestones* that operators must achieve in order to drive on public roadways.

- *Milestone 1: Ability to safely conduct testing of autonomous vehicles with a safety driver in a small-scale testbed.*

**Fig. 1.** Singapore automated shuttle [11]

- *Milestone 2: Ability to safely conduct testing of autonomous vehicles with a safety driver in a complex environment.*
- *Milestone 3: Ability to safely conduct testing of autonomous vehicles without or with a safety driver (with limited control) in a complex environment. This implies high technical maturity* [11].

In 2019, Singapore conducted its first AV shuttle pilot as shown in Fig. 1 and has expanded AV pilots since then, working with academic partners and closely monitoring the AVs involved in the pilots [12].

According to a report from KPMG when it comes to AV Readiness, the United States has fallen behind Singapore, The Netherlands, and Norway in several categories. The US's scores highest in the technology and innovation category. However, when it comes to policy and legislation, the US ranks 6$^{th}$. Singapore ranks first overall and first for policy and legislation, resulting from the government's national standards and an AI governance framework. The country also ranks first for consumer acceptance, which may be related to the stringent standards and regulation put in place. The UK ranks second for policy and legislation, resulting from the Automated and Electric Vehicles Act of 2018, and the inclusion of the regulations discussed in the WEF report [13].

## 2.2 Reforms Proposed to Federal Motor Vehicle Safety Standards

There are a range of perspectives from government and private-sector stakeholders on how best to regulate AVs to promote safety and efficiency, while encouraging innovation. Despite the wide range of perspectives, there is agreement on the key policy issues in need of reform.

The USDOT has expressed caution for developing national standards in the nascent AV industry in order to encourage innovation. The government claims that there are many unknowns in the industry. However, the lack of durable technical standards at the federal level makes it difficult for vehicle manufacturers and ADS designers to operate in multiple U.S. states, and adds risk for companies as they develop their long-term investment strategy. Critics of the FMVSS rulemaking reform process lament that rules can take six to eight years to complete. This pacing problem may pose an ongoing challenge for the development of AV and ADS regulations, with some arguing it could put U.S. AV companies at a competitive disadvantage as international standards emerge more rapidly. Others see the long regulatory process as a necessary part of ensuring fairness in the process.

In light of the historic timeframes for establishing standards, and regulatory incorporation of those standards, some experts suggest that Congress and NHTSA should reevaluate the existing FMVSS exemption regime. Previously introduced, but not enacted, AV legislation proposed to increase the annual exemption cap from 2,500 vehicles to between 80,000 and 100,000 in recognition that automated vehicle development and deployment is likely to outpace the development of safety and performance regulations. Among industry and government there is some agreement that the cap should be re-evaluated and the existing equivalent safety requirement maintained. There are also some who express concern that as the number of exemptions grow, so does the public risk burden, with more noncompliant AVs operating on the roadways. As NHTSA considers restructuring FMVSS exemptions with the safety framework approach (see Sect. 2.1) in the December 2020 ANPRM, there are still questions about whether the agency possesses the technical understanding to properly evaluate automated vehicle exemption petitions, VSSAs, or to develop a more robust performance-based safety framework for the ADS software.

### 2.2.1 Considerations for Federal Preemption

In the absence of federal regulation specific to automated vehicles, some states are taking policy actions that might otherwise be federal in nature. This has resulted in a disjointed state and local AV legal landscape that may be undesirable for several reasons. First, there may be a lack of technical expertise on AVS within some state agencies, and NHTSA may be better equipped to set standards (although as previously mentioned, NHTSA may also be lacking in certain technical expertise areas). Second, in order to achieve continuity across states, federal preemption of states on AV policy matters may be necessary. But there is disagreement among federal policymakers about where the boundaries of preemption should begin and end. A tailored preemption of state automated vehicle safety regulations could be understood as preserving longstanding divisions of authority rather than upending the auto safety regulatory ecosystem. During the 2018 deliberations surrounding a federal AV bill drafted in 2017, there was discussion about using the term AV "performance" in the preemptive language. This type of term is difficult to interpret, and could result in wide boundaries that could interfere with state and local AV traffic control and environmental regulatory purview [14]. Future legislation targeting AVs will be best served to establish clearer boundaries limiting federal purview or demonstrating a clear rationale for where uniformity across state and local borders serves the public interest.

### 2.2.2 Considerations for Data Collection

More federal research on AV regulation is needed. Research needs fall into a number of categories, including research into how best to execute research. These categories include identifying the types of data collection and analysis that operators can apply to various use cases. Some argue that the AV sector should strive to achieve some international harmonization for data collection. The lack of comparable data on automated driving presents challenges to future regulation.

USDOT has been involved in several initiatives that suggest an interest in data collection and sharing around AVs. The agency developed the Data for Automated Vehicle Integration (DAVI) initiative "to identify, prioritize, monitor, and – where necessary – address data exchange needs for automated vehicles (AV) integration across the modes of transportation" [15]. Although DAVI is still in the early stages, it has outlined several goals including promoting data-driven safety and enabling voluntary data exchanges as shown in Fig. 2.

DAVI has yet to accomplish these goals and is still in its early stages. Future USDOT leadership could choose to develop DAVI further or choose to pursue other avenues for data collection and analysis. For data collection, there is a public sector-led route and a private-sector led route, among other options. But the best data collection and sharing platforms rely on clear partnerships between these sectors, and consider data privacy to be a top priority [15].

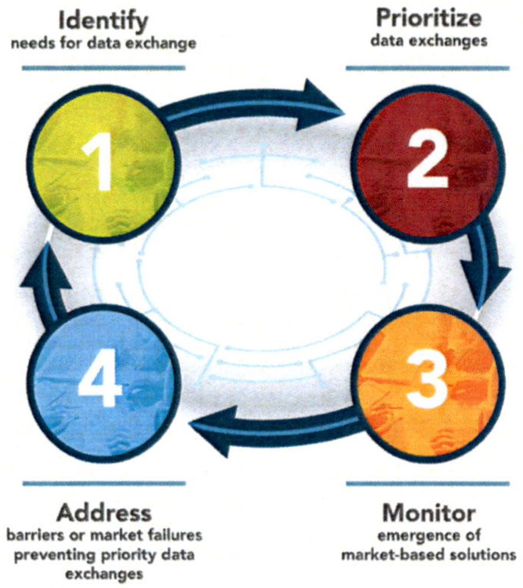


**Fig. 2.** Data for Automated Vehicle Integration data exchanges

There are several examples of public-sector data sharing platforms capable of housing big datasets. An exemplary model is the USDOT's Secure Data Commons (SDC). The SDC is a third-party platform for storing and analyzing potentially sensitive mobility data. This type of data which includes either personally identifiable data (e.g. license plates) or data that can be re-identified if there is enough historic data (e.g. months of specific routes including GPS coordinates). The SDC system addresses the potential for data privacy issues, by hosting two main types of users: data providers and data analysts. Data providers voluntarily add data; even real-time data can be hosted by this platform. Providers can also upload historic datasets on an occasional or ad-hoc basis. The SDC grants different levels of access to different types of data analysts based on criteria defined by data providers. Providers also set parameters for whether datasets (or data analyses conducted on the SDC platform) can be exported. Data analysts have access to a set of datasets tailored specifically to the level of access they are granted, as well as to statistical tools such as R, Python, and SQL. This type of data exchange ensures that large datasets can be shared with the public sector without full public disclosure, which improves security for data providers and preserves user privacy [16, 17].

Another way to share data and to encourage industry-wide learning, is to establish business-to-business (B2B) data sharing platforms such as the Project for the Establishment of Generally Accepted quality criteria, tools and methods as well as Scenarios and Situations (PEGASUS) which was launched with support from the German Federal Ministry for Economic Affairs and Energy (BMWi) and focuses on highly automated vehicles (HAVs). The platform enables sharing of HAV test scenario designs, which allows operators to learn from the testing efforts of other businesses [18]. Supporting, or even requiring participation, in this type of B2B information exchange could improve safety across the industry, by enabling companies to learn from the testing experiences of their peers in a more meaningful way than simply records of collision or near-collision data, which are not very useful for encouraging continuous improvement across the industry.

While these types of exchanges could host or encourage sharing of AV data, NHTSA and the Federal Motor Carrier Safety Administration (FMCSA) have also indicated they are undertaking research related to operator requirements for automated trucks. FMCSA requires commercial AV operators to be subject to truck-specific requirements—that are unique to trucks, including requirements that are not directly related to the driving task, such as ensuring cargo is properly secured—and that special attention should be given to FMCSA. These unique requirements may point to a need for a dedicated channel of data collection or data sharing for truck heavy duty AV operators and regulators [5].

### 2.2.3 Considerations for Shared Passenger Fleet Operation

Safety performance metrics for shared passenger fleet operations need to be different from metrics for personally owned passenger AVs. States and localities currently have the legal authority to regulate the operation of for-hire services (e.g. taxis, charter buses, ride-hailing operators). Federal guidance on these operations could ensure more continuity across the industry, but state purview is well-established. More discussion is necessary to determine if there are additional safety performance measures that NHTSA could enact to incentivize vehicle design in such a way that shared-use is more appealing. Encouraging shared use could aid in meeting state emissions and traffic reduction goals.

Reforms to the FMVSS could encourage more safety features for shared passenger travel in AVs. There may be a federal role in encouraging design features that make sharing space more comfortable and secure. According to UC Davis researchers, "Seating configuration and territorial props in robo-taxis can influence perceived personal space" and these factors can encourage or discourage pooling in AVs (see Fig. 3 for more on this research into AV design) [19]. These FMVSS reforms could also include considerations for ensuring the security of passengers aboard shared AVs, such as a requirement

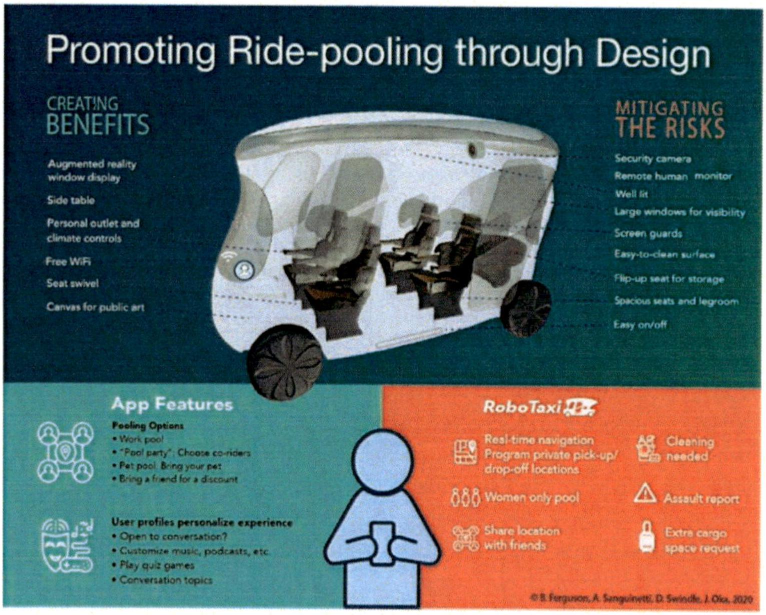


**Fig. 3.** AV design to promote shared rides.

for ensuring remote or physical intervention protocols for use in the event of passenger dispute or assault.

## 2.3 Connected Vehicle Technology, the FCC and Advancing Automated Vehicle Safety

For nearly two decades state departments of transportation (DOTs), local governments, and other organizations have been working diligently to improve highway safety by leveraging emerging technologies. Many state and local governments believe that automated vehicles need redundant technologies, including connected vehicle (CV) technologies to ensure that automated vehicles truly advance highway and roadway safety. Connected vehicle technologies allow vehicles to directly communicate with each other, roadside **infrastructure, other road users. These technologies can advance road safety, efficiency,** smart mobility, and sustainability [20].

One such CV technology is dedicated short-range communications (DSRC) - a radio technology that has been tested and deployed by DOTs and local governments - to advance highway safety. Radio technology regulations fall under the Federal Communications Commission (FCC) oversight. In November 2020 the FCC voted to approve a new rule that re-allocated portions of the radio spectrum away from DSRC technology to allow private industry to advance connected vehicle technologies that can communicate with anything, often called "cellular vehicle to anything" technology, or "C-V2X" including 5G and other Wi-Fi technologies [21].

Historically, the FCC preserved specific parts of radio bandwidth for safety technology, commonly referred to as the "safety spectrum" to promote transportation safety. Emergency providers, law enforcement, safety advocates, and transportation officials across the country have used this spectrum to advance transportation technologies that save lives, most recently evidenced by the "signal phasing and timing (SPaT) challenge" advanced through the American Association of State Highway and Transportation Officials (AASHTO). This challenge encouraged state DOTs to install roadside units that use DSRC technology on traffic signals to communicate basic safety messages to vehicles – all of which required use of the safety spectrum [22]. These basic safety messages allow traffic signals to share their phasing timing with connected vehicles, which allows drivers to know when a light may turn, communicate latitude and longitude and speed of the vehicle to the traffic signal, and advance eco-driving by encouraging vehicles to operate more efficiently to move through intersections. Many states successfully deployed DSRC "connected" corridors using DSRC technology.

However, similar to many emerging technologies, there are private industry competitors trying to find the "best" solution to transportation safety and efficiency. A competing technology – cellular technology – is being developed by the telecommunications industry to increase transportation safety. The C-V2X cellular technology has not been fully tested in the U.S.

The FCC noted in its ruling and comments that state DOTs and the DSRC industry innovation stagnated, with zero-to-few vehicles incorporating DSRC technologies. C-V2X industry members and automobile manufacturers advocated that new technologies like wi-fi need to be integrated into vehicles to preserve safety, similar to many of the cellular solutions already available on the market in other sectors. To encourage the FCC to release the radio spectrum to allow for cellular technology solutions, certain automobile manufacturers committed to integrating C-V2X technologies by 2022 including Volvo, Audi, BMW, Daimler, and Volkswagen.

The USDOT and state DOTs generally advocate "technology neutrality", which means that they do not advocate one technology over another. However, the USDOT and other advocacy groups such as AASHTO and the Intelligent transportation Society of America have opposed the new FCC ruling on the grounds that these newer technologies have not been fully tested and could lead to dropped signals, as is common in some areas with cell phones. Despite opposition, the FCC rule was adopted. Currently, state DOTs, local governments, and advocacy groups that support DSRC technology are seeking resolution of the ongoing issue, including considering requesting compensation for investing in DSRC technologies that are now obsolete. The FCC is allowing governments time to discontinue these technologies and is not approving DSRC licensing. Meanwhile the DSRC industry has quickly discontinued supporting these technologies that have already been deployed.

While connected vehicle technologies will continue to advance and evolve, it is critical to understand industry adoption of new technologies before states and local governments fully integrate these technologies. Governments must remain nimble and flexible when adopting CAV technologies.

## 3 State Automated Vehicles Policy and Regulation

The limited role that the Federal government has played thus far in AV regulation has led states to develop more comprehensive design and implementation policies. However, the scope and intention of these policies varies between states. Some states have taken a permissive approach, allowing AV companies to test and operate in their state broadly, while others have taken a more restrictive approach, requiring licensing, permitting, and other restrictions.

State governments oversee driver licensing, vehicle registration, insurance requirements, vehicle inspections, and traffic laws to ensure safety. States also play a significant role in funding and building transportation projects. There is considerable variability in how states have exercised these powers with respect to AVs [23].

The focus of AV policies is also broad, as states have varying concerns. A recent issue paper from UC Davis analyzes the database from the National Conference of State Legislatures (NCSL), finding trends throughout state policies. AV policies include executive orders and legislation (regulatory actions were not tracked). The policies in the NCSL database vary based on the level of automation being regulated, and how they impact society. A majority of the policies focus on safety, but some policies address environmental, equity, and privacy goals [24] (see Fig. 4).

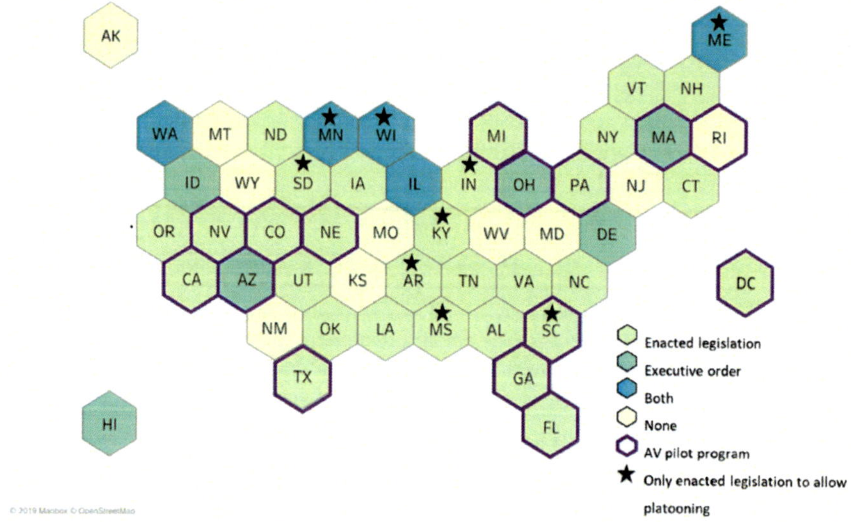


**Fig. 4.** AV policies enacted by state

Policies enacted by each state as of April 2020. States outlined in purple have state-implemented AV pilot programs. States in light yellow do not have any AV policies, states in light green have enacted legislation, states in green have enacted an executive order, and states in blue have enacted both. States denoted with stars have only enacted policies related to platooning.

The most common policies across the US are related to platooning vehicles, mainly in amending or exempting AVs from laws that prohibit following too closely. This policy focus is likely because long-haul trucking is an attractive application of AV technology because trucks drive on long stretches of highways at consistent speeds, and can utilize platooning to save energy [25]. Other policies to date are still preliminary in scope. They either a) codify the definition of AV technologies, or b) assemble groups (e.g. committees, working groups, or task forces) to investigate the societal, environmental, and safety impacts of AVs. These impacts are unique to each region, even within states, so preemption should be carefully considered at the state and federal level so that local governments can design policies to fit their unique needs and goals [24].

States are also beginning to designate regulatory policies and regulatory bodies. The UC Davis study found that most states that have enacted regulatory policies have designated a regulatory body, and many have designated the regulatory body before enacting any regulations. Many states have multiple agencies in charge of regulation for different policies, including licensing and permitting, environmental impacts, business operations, and safety. This multiple-agency strategy may make AV policy regulation more complex, but it may also ensure that all impacts are carefully considered by agencies with the technical expertise to address the different aspects of AV policy [24].

### 3.1 AV Liability and Insurance Policy Considerations

States have historically led on liability and insurance policy, and will likely remain a key player in developing policies that assign liability for AVs, although the courts will likely play a role. However, liability theory will likely require updates to implement where the AV occupants have no vehicle control. Liability can be reassigned to AV manufacturers, although this might have a chilling effect on the industry because manufacturers would have to assume all risks in an emerging industry. This manufacturer led liability scheme may be appropriate for crashes caused by flaws in AV hardware or software design, which are expected to occur less frequently than current crashes but will occur nonetheless. There may be other actors that will be assigned liability in some crashes. New mobility companies may own AV fleets to operate them in taxi services. Fleet owners may also contract with transportation network companies (TNCs) like Uber and Lyft to dispatch AVs using apps and charge rates based on travel demand algorithms rather than taxi rates. These entities will likely assume some of the crash liability and all of the liability for security risks that might occur within the vehicle [26].

As a UC Davis policy research report states, it may be beneficial for state lawmakers to act proactively to address these issues and distribute liability clearly within the industry.

> *If lawmakers cannot establish a comprehensive AV governance policy, AV liability policy will be instead be set by the judicial system. This is far from ideal. Leaving AV liability policy up to individual judges would exacerbate and extend inconsistency and uncertainty. This would make it difficult and expensive for victims of AV crashes to recover damages* [26].

## 3.2 Highlighting AV Regulatory Landscape in Select U.S. States

### 3.2.1 Arizona AV Policy

Unlike other states with robust automated vehicle testing taking place on public roads, until recently Arizona lacked legislative and regulatory frameworks. Instead, testing in Arizona was underpinned by three executive orders. The first in 2015 ordered state agencies to prepare for and facilitate on-road automated vehicle testing and deployment, established university-based pilot programs, and created an oversight group within the governor's office [27]. The second in early 2018 updated definitions to conform to consensus standards and explicitly stated that automated vehicle operators in the state must comply with all applicable state and federal laws and regulations. This executive order also established the need for a "minimal risk condition" in the event the AV fails to operate, and allowed the state to revoke permission to operate in the state if these conditions are not met [28].

**Fig. 5.** Member chart of the Institute of Automated Mobility coalition, established in Arizona.

The third in late 2018 created the Institute of Automated Mobility (IAM) within the Arizona Commerce Authority to coordinate technical and policy research activities within the state [29]. The Institute established a coalition (as show in in Fig. 5) and identified their role in addressing the following problem, "The large-scale market and societal disruption being caused by automated vehicles demands industry-driven validation of consistent safety standards, policies and technology neutral solutions that don't yet exist" [30].

In March 2021, Arizona's governor signed into law House Bill 2813, which formally codified Arizona's AV testing framework that had been established by executive order.

### 3.2.2 California AV Policy

California's AV regulatory program includes considerable oversight from several different agencies. The resulting rules and regulations reflect a compromise between cities, advocates, and companies. Vehicle safety is regulated by the California Department of Motor Vehicles (DMV). Passenger service is regulated by the California Public Utilities

Commission, given their historic role in overseeing statewide taxi and chartering passenger service (e.g. employee shuttles, limousines). Lastly, the California Air Resources Board is tasked with developing emissions regulations for AVs used in commercial passenger service (although this program will be administered by the CPUC).

*California's Vehicle Safety and Testing Regime*
The CA DMV regulates AV design by issuing testing permits that allow on-road testing of AVs and AV technology. The DMV requires AV companies to submit a self-certification that is voluntary under NHTSA rules. Regulatory authority was established within the DMV through passage of California State Senate Bill (SB) 1298, requiring the California Department of Motor Vehicles (DMV) to develop rules for AV operation [31].

The resulting regulation went into effect in April 2018. California's policy requires companies seeking to test Avs to apply for permits from the State. California also allows for "remote" testing without a driver or operator in the vehicle so long as a permit has been obtained and a law enforcement interaction plan has been submitted. Finally, California requires AV testing companies operating in California to document and report miles driven as well as the number of times when a human driver, when present, had to retake control of the vehicle [32].

*California's AV Passenger Service and Shared Mobility Applications*
The California Public Utilities Commission (CPUC) moved the state's AV Program for passenger service from pilot to deployment in a decision on November 19, 2020. This decision maintained the state's authority to regulate AV's in passenger service and denied requests from cities to require local approval for AVs operating in their jurisdictions. The decision established four goals for AV operation in California [33]:

1) Protect passenger safety;
2) Expand the benefits of AV technologies to all of Californians, including people with disabilities;
3) Improve transportation options for all, particularly for disadvantaged communities and low-income communities; and
4) Reduce greenhouse gas emissions, criteria air pollutants, and toxic air contaminants, particularly in disadvantaged communities [33].

These goals establish important guardrails for the state regulatory framework for Avs. Other considerations related to the November 2020 CPUC rule:

- **This decision lifts a previous ban on pooling or sharing rides between parties in a single AV:** Allowing pooling was a noncontroversial decision for the CPUC. There was a consensus from the more than 20 parties to the CPUC proceeding, in comments submitted to the Commission, stating that pooling is necessary for Avs in order to meet traffic and sustainability goals. A large body of research supports the claim that sharing rides will be a critical part of ensuring affordability and sustainability of Avs. For these reasons it may not be enough for policy makers and practitioners to tolerate, or allow, pooling, but they may need to champion pooling [33].
- **The decision lifts a ban on compensation for all rides.** Allowing compensation was also widely supported by all but a few parties to the proceeding. The notable exceptions

were the San Francisco Municipal Transportation Authority, San Francisco advocates for a slower path from pilot to full deployment, in order to give cities more control over—and time to prepare for—the types and numbers of Avs allowed on its streets. San Francisco believes that restricting compensation will slow the growth of the AV service industry [33].

- **The CPUC will require AV companies to submit reports that ensure that safety measures apply to all passengers, including those with disabilities.** However, CPUC will not require that AV operators provide equivalent accessible service for people with physical or mental disabilities. This type of safety guarantee does not promise universal design for all AVs, which is what disability advocates preferred in their comments to the proceeding. Further, while the safety needs of disabled passengers are critical, this decision stops short of ensuring that information disbursal and payment processing meets the needs of riders with disabilities [33].

*California AV Data Sharing Requirements*
There are currently 58 AV companies that have historically been approved for testing with a driver, and five approved for driverless testing. One company (Nuro) has received a commercial deployment permit. For all these permit holders DMV requires collision reports, and reports on miles traveled and number of vehicles in operation. According to DMV, "Manufacturers need to report any collision that results in property damage, bodily injury, or death within 10 days of the incident" [32]. It is not clear if the collision reporting metrics are useful for determining the safety of the AVs operating in California, given that collisions align with miles driven, and the simplistic metric is not particularly valuable in determining how to prevent similar type collisions from occurring in the future.

The CPUC requires companies to provide publicly available quarterly data reports (although they can request some/all data be redacted). These data reports include each AV trip's time, date, census tract, and occupancy, among other metrics. Reporting includes total empty miles driven, times when neither a driver nor a passenger is in the vehicle, as well as detailed information about the number of electric vehicles in the fleet and their charging behavior [33].

## 4 Regional Coalition-Building and Efforts

Several state coalitions are leading in developing goals, testing procedures, and shared protocols across U.S. state borders. These activities speak to the need for conformity across borders and also the benefits that states can incur by working collaboratively.

### 4.1 Eastern Transportation Coalition I-95

The Eastern Transportation Coalition (formerly the I-95 Corridor Coalition, now referred to as TETC) is a partnership of state Departments of Transportation and related authorities and organizations that work together to accelerate improvements in freight and passenger movement along the eastern seaboard. As a component of its Innovation Program Track, the Coalition's Connected and Automated Vehicle (CAV) Program supports members as they address new and emerging transportation challenges in new technologies, policies and multi-state partnerships. This work helps staff at member agencies identify the practical actions they can take and implement the technology of tomorrow today [34].

The CAV program activities are designed to respond to the needs of TETC's members. In September 2020, TETC hosted a webinar on State DOTs and communications about CAV [35]. In late 2020, TETC worked on a USDOT/FHWA funded project to create a regional readiness assessment for automated vehicles. This project was tested by three pilot states and was completed in January 2021. The Coalition is working to collaborate with other AV readiness and highway automation projects and is working with members to explore next steps [34].

### 4.2 Smart Belt Coalition PA-OH-MI

The Smartbelt Coalition includes several agencies and academic institutions including the Ohio Department of Transportation (ODOT), the Ohio Turnpike and Infrastructure Commission, The Ohio State University and Transportation Research Center, The Pennsylvania Department of Transportation (PennDOT), Pennsylvania Turnpike Commission (PTC), Carnegie Mellon University, the Michigan Department of Transportation (MDOT), and the University of Michigan [36].

In April 2020, the Smart Belt Coalition issued a request for information regarding truck automation and platooning. The coalition is seeking a demonstration of truck platooning that would cross all three states in the belt. Platooning is the most popular state AV policy type and aligns well with the issues many states are confronting. Notably, Pennsylvania law requires that platooning trucks include a visual identifier that can signal to law enforcement and other road users that a platooning relationship is occurring. This is an example of where coalitions can result in similar policy in neighboring states [37].

## 5 Conclusion

The U.S. is a fragmented and asynchronous AV policy landscape that adds regulatory hurdles to an industry which is tasked with tackling significant technological hurdles. In order to realize the potential benefits of AVs rapidly, innovators need a durable and balanced policy regime. This system will need to identify a balance between regulatory guardrails and overly prescriptive oversight. We can learn from the successes of other countries, including Singapore and the UK, as well as looking towards the nation's states as policy laboratories. This paper aims to address the U.S. Federal and State Policy landscape for AVs by summarizing and expanding on our session, Regulatory *Policy for Automated Vehicles*' at the Automated Vehicle Symposium in July 2020. This paper presents a balanced account that will identify key issues, tradeoffs and actions for AV policy. There are opportunities for future research that gathers the policy alternatives listed in this section and develops a more rigorous method for measuring their effectiveness and confronting the tradeoffs.

## References


1. NHTSA: Framework for Automated Driving System Safety. U.S. Department of Transportation, Washington, D.C. (2020)
2. NHTSA: Federal automated vehicles policy. US Department of Transportation (2016). https://www.transportation.gov/sites/dot.gov/files/docs/AV%20policy%20guidance%20PDF.pdf
3. NHTSA: Automated Driving Systems: A Vision for Safety 2.0. US Department of Transportation (2017)
4. US Department of Transportation: Preparing for the Future of Transportation: Automated Vehicles 3.0. https://www.transportation.gov/sites/dot.gov/files/docs/policy-initiatives/automated-vehicles/320711/preparing-future-transportation-automated-vehicle-30.pdfPreparing
5. Federal Motor Carrier Administration: Safe integration of automated driving systems-equipped commercial motor vehicles; correction. US Department of Transportation (84 FR 25229) (2019). https://www.govinfo.gov/content/pkg/FR-2019-05-31/pdf/2019-11387.pdf
6. Federal Railroad Administration: Automated vehicles at highway-rail grade crossings. Final Report. U.S. Department of Transportation (2018)
7. Federal Transit Administration and John, A.: Volpe national transportation systems center. Strategic Transit Automation Research Plan, US Department of Transportation, vol. 0116 (2018). https://www.transit.dot.gov/sites/fta.dot.gov/files/docs/research-innovation/114661/strategic-transit-automation-research-report-no-0116_0.pdf
8. Federal Transit Administration: Transit Bus Automation Quarterly Update. U.S. Department of Transportation (2020)
9. Maritime Administration: Request for information on opportunities, challenges and impacts of automated transportation in a port environment. U.S. Department of Transportation (84 FR 37951) (2019)
10. Avary, M., Dawkins, T.: Safe Drive Initiative: creating safe autonomous vehicle policy. In: World Economic Forum (2020). http://www3.weforum.org/docs/WEF_SafeDI_creating_safe_AV_policy_2020.pdf
11. de Boer, N.: Singapore TR 68 : 2019 - Technical Reference for Autonomous Vehicles, Singapore (2019). https://cdn-advi.s3.ap-southeast-2.amazonaws.com/wp-content/uploads/2019/11/1100-Niels-de-Boer-Singapore-TR-68-2019-Technical-Reference-for-autonomous-vehicles.pdf
12. Land Transport Authority: Autonomous vehicle testbed to be expanded to western Singapore – continued emphasis on public safety. Singapore Government Agency (2019)
13. 2019 Autonomous Vehicles Readiness Index - KPMG Global. https://home.kpmg/xx/en/home/insights/2019/02/2019-autonomous-vehicles-readiness-index.html. Accessed 17 Mar 2020
14. Davis, J.: New AV START discussion draft addresses some stakeholder concerns. Eno Center for Transportation, 05 December 2018. https://www.enotrans.org/article/new-av-start-discussion-draft-addresses-some-stakeholder-concerns/
15. U.S. Department of Transportation: Data for Automated Vehicle Integration (DAVI), 04 December 2019. https://www7.transportation.gov/av/data
16. Cohen D'Agostino, M., Pellaton, P., Brown, A.: Data sharing: challenges and policy recommendations. UC Davis: Policy Institute for Energy, Environment, and the Economy (2019). https://escholarship.org/uc/item/4gw8g9ms
17. U.S. Department of Transportation: What is SDC? 06 February 2020. https://www.transportation.gov/data/secure/about
18. Pegasus: Pegasus Research Project Securing Automated Driving Effectively. https://www.pegasusprojekt.de/en/about-PEGASUS

19. Sanguinetti, A., Ferguson, B., Oka, J., Alston-Stepnitz, E., Kurani, K.: Designing Robo-taxis to promote ride-pooling. UC Davis (2020). https://escholarship.org/uc/item/65s3m92w
20. White, K., Huebsch, C.: Minnesota Fiber Optic Feasibility & Partnership Study (2020). http://www.dot.state.mn.us/automated/docs/fiber-optic-feasibility-and-partnership-study.pdf
21. **Federal Communications Commission: FCC modernizes 5.9 GHz band for wi-fi and auto** safety, 18 November 2020
22. FCC Opens 5.9 GHz Spectrum to Non-Transportation Use. AASHTO Journal. https://aashtojournal.org/2020/11/20/fcc-opens-5-9-ghz-spectrum-to-non-transportation-use/. Accessed 29 Dec 2020
23. Anderson, G., Brown, A.L., Safford, H.R.: The road to successful governance of automated vehicles. Policy Brief. Policy Institute for Energy, Environment, and the Economy (2018). https://epm.ucdavis.edu/sites/g/files/dgvnsk296/files/inline-files/AV%20PolicyBrief_FINAL_TH_2018_17_12.pdf
24. Fleming, K.: Technology is outpacing state automated vehicle policy. Policy Institute for Energy, Environment, and the Economy (2020). https://escholarship.org/uc/item/0k85r9jv
25. **US Department of Energy: Platooning trucks to cut cost and improve efficiency.** Energy.gov. https://www.energy.gov/eere/articles/platooning-trucks-cut-cost-and-improve-efficiency. Accessed 21 Dec 2020
26. Anderson, G., Brown, A., Safford, H.: Policy Brief: reshaping liability and insurance rules for automated vehicles. 3 Revolutions. University of California, Davis, 18 Jun 2019. https://3rev.ucdavis.edu/policy-brief/reshaping-liability-and-insurance-rules-automated-vehicles. Accessed 10 Dec 2019
27. Ducey, G.D.: Executive Order 2015-09: self-driving vehicle testing and piloting in the state of Arizona; self-driving vehicle oversight committee (2015)
28. Ducey, G.D.: Executive Order: Advancing Autonomous Driving Testing and Operating; Prioritizing Public Safety (2015)
29. Ducey, G.D.: Executive Order: Establishment of the Institute of Automated Mobility (2015)
30. Institute for Automated Mobility Shaping the Future of Transportation, Safety, Science and Policy. https://webadmin.azmag.gov/Portals/0/Documents/MagContent/EDC_2019-04-2_Item-5-Institute-for-Automated-Mobility.pdf?ver=2019-04-02-094836-437
31. Padilla: Vehicles: autonomous vehicles: safety and performance requirements, vol. Division 16.6 Section 38750 (2012)
32. State of California, Testing of Autonomous Vehicles - Adopted Regulatory Text, vol. Title 13, p. 32 (2021)
33. California Public Utilities Commission. Decision Authorizing Deployment of Drivered and Driverless Autonomous Vehicle Passenger Service (2020)
34. The Eastern Transportation Coalition. FY2020 Year in Review (2020)
35. Reeder, V.: Communicating CAV for DOTs: Public Perception, Awareness, and Education. The Eastern Transportation Coalition (2020). https://tetcoalition.org/wp-content/uploads/2020/10/September-24-CAV-Webinar-Slides-FINAL.pdf
36. Smart Belt Coalition Strategic Plan (2017)
37. Smart Belt Coalition: Driver-Assistive Vehicle Platooning. Request for Information (2020). http://www.dot.state.oh.us/Divisions/ContractAdmin/Contracts/PurchDocs/606-20.pdf

# Regulation of In-Service Safety Risks of Automated Vehicles

Marcus Burke(✉)

Future Technologies, National Transport Commission,
Level 3, 600 Bourke Street, Melbourne, Australia
mburke@ntc.gov.au

**Abstract.** Automated vehicles come with a range of safety risks, that could result in deaths or serious injuries on the road. Companies will seek to address these risks in the initial design of the automated driving system. However, automated vehicles will need to operate safely not just on day one, but throughout their lifetime on the road. On-road or "in-service" safety risks could result from degradation of the vehicle or changes to the outside environment. Companies developing automated driving systems will need to manage these ongoing safety risks; government will need to ensure appropriate regulation of these risks.

**Keywords:** Automated vehicle · Safety · In-service safety · General safety duties · Safety management

## 1 Automated Vehicles and Regulation

Automated vehicles are vehicles that include an automated driving system (ADS) that is capable of monitoring the driving environment and controlling the dynamic driving task (steering, acceleration, and braking) with limited or no human input [1]. Automated vehicles offer significant safety, productivity, environmental and mobility benefits.

However, there are a range of uncertainties around the potential deployment of automated vehicles, including:

- Likely timing and extent of deployment
- Applications and operational design domains
- The mix of technologies that will be used and
- How they will change the behaviour of road users.

It is anticipated that automated vehicles will be safer than human driven vehicles, since they will never break the law, never speed, get distracted or drive while fatigued and have 360-degree vision of the road.

There will still be safety risks associated with automated vehicles though. Automated vehicles could crash for a range of new reasons, including poor design, inaccurate or out-of-date information inputs and user confusion, in addition to existing risks such as weather and damaged infrastructure. Automated vehicles are not currently legal in many


G. Meyer and S. Beiker (Eds.): AVS 2020, LNMOB, pp. 40–46, 2022.
https://doi.org/10.1007/978-3-030-80063-5_4

countries, where laws are premised on the presence of a human driver. A key challenge for governments is how to update our current regulatory systems to ensure that people can gain the benefits of this technology, whilst also managing the risks. That is, how to remove the barriers in legislation whilst also addressing the gaps [2].

## 2 In-Service Safety Risks and Automated Vehicles

Ultimately the key safety risk for automated vehicles is that they crash or collide causing death, serious injury and/or property damage.

This could be the result of a variety of causes including:

- Poor system design
- Faulty sensors
- Failure to predict the movements of other road users
- Failure to account for local conditions, such as different road rules
- Use of incorrect information, such as inaccurate speed zone information
- User errors such mode confusion or failure to respond to requests to take over control.

There may also be safety factors which apply to all vehicles, including:

- Dangerous weather conditions, such as ice on the road
- Dangerous behaviour by other road users
- Poor quality or damaged infrastructure (such as potholes in the road)

A clear focus to date has been on assessing the safety of these vehicles at first market entry. That is, how to assess the safety of new automated vehicles, before they are allowed to operate on public roads. What standards or criteria should be used to assess their safety? What evidence or metrics can be used to demonstrate or measure their safety?

In Australia, transport ministers have agreed a self-certification approach under which companies proposing to bring automated driving systems to market will need to provided evidence against eleven safety criteria [3]:

1. Safe system design and validation processes
2. Operational design domain
3. Human-machine interface
4. Compliance with relevant road traffic laws
5. Interaction with enforcement and other emergency services
6. Minimal risk condition
7. On-road behavioural competency
8. Installation of system upgrades
9. Verifying for the Australian road environment
10. Cybersecurity
11. Education and training

These criteria are largely focused on the *design* of the system, but also take in account processes for managing change and processes that the organization must have in place and manage. However, governments and the public will want to know not just that vehicles are safe on day one, when they enter the market, but that they continue to operate safely throughout their operational life.

Automated driving systems will not be a static product that companies can send out into the world. They will be dynamic systems operating in a dynamic environment. In many ways they are moving from a product, in the form of a traditional vehicle for sale, to a service (automated driving) – a service that will need to be maintained over time. The external environment will not be static and hence automated driving systems cannot be static.

Automated vehicles may have shorter lifespans on the road than current vehicles, particularly in they can are driven many more hours of the day than current vehicles. However, even if vehicles are only in-service for three to six years, rather than current vehicles (which can operate for 15 years or more) this represents a substantial period during which they must continue to operate safely.

Automated driving systems will need to be monitored and updated to take into account changes in the external environment, including:

- changes to infrastructure, such as new roads or new styles of road signs
- changes to road rules and other laws
- changes to cybersecurity threats
- changes to the mix of vehicles on the road, for example the arrival of new vehicle types such as e-scooters.

Monitoring must also include monitoring for deterioration of the hardware of the vehicle, particularly the vehicle sensors.

Any in-service safety regime must take these safety risks into account. Without ongoing monitoring and support, an automated vehicle can become a danger to its occupants and to other road users, particularly to vulnerable road users. There is a clear case to consider government action.

## 3 In-Service Regulation of Automated Vehicles

The regulation reform program for the commercial deployment of automated vehicles in Australia has sought to answer five key questions:

1. Who is in control of an automated vehicle?
2. How do we ensure automated vehicles are safe when they first enter the market?
3. How do we ensure automated vehicles operate safely throughout their life on the road?
4. How do we manage injury insurance for automated vehicles?
5. How do we manage access to data?

For this chapter, we will focus on question 3, How do we ensure automated vehicles operate safely throughout their life on the road?

A key issue is to examine the in-service safety risks of automated vehicles (as set out above) and then to look at how these risks are covered by regulation today. A number of legal parties in relation to the automated vehicle are already covered by existing legislation today – for example human drivers, vehicle owners and vehicle repairers. The current regulations could potentially cover the new risks of automated vehicles for these parties. For example, obligations on repairers could extend to the safe repair of automated driving systems [4].

However, the companies bringing the technology to market, what in Australia we have referred to as the Automated Driving System Entity (ADSE) is not covered by existing legislation. So, there is a potential gap in coverage of the party with the most responsibility – the company bringing the automated driving system to market and maintaining it to ensure it continues to operate safely.

Even if there is a potential gap, a question arises as to the need for government intervention – can the risks simply be left for the market to manage? Whilst there will be a strong commercial imperative to develop safe products – consumers will not likely buy or pay to use automated vehicles that are unsafe – there is a risk that consumers will be unaware of safety risks until it is too late. There is also a risk that poor operators bring automated vehicles into the market without the necessary means to support their ongoing safe operation. Crashes caused by unsafe operators could undermine the entire industry. And there is a strong argument that companies that produce poor systems that result in deaths and serious injuries should face a stronger remedy than just loss of potential future profits.

These risks are partially addressed through regulation at first supply. For example, ensuring that there has been a safe system design and validation process. But this will not ensure that vehicles are maintained in way to ensure that they operate safely, again taking into account the dynamic external environment.

Regulation can also benefit the safe operators, by keeping the unsafe operators who may undermine confidence in the whole industry – out of the market.

However, if we have established the need for government intervention and regulation, a further question remains as to *how* to regulate. Should governments attempt to identify and manage every risk posed by an automated driving system? Could they foresee every scenario and provide rules for the range of different technologies, applications, vehicles, and business models that are likely to evolve? Or are there alternative ways to approach the problem?

## 4 In-Service Safety and General Duties

An alternative approach for governments is to regulate for the outcome – safety – rather than trying to prescribe the means to achieve that outcome or the means to address each individual risk.

An approach that has been successful in other areas of safety regulation is to set out general safety duties for responsible parties. This has been used in Australia to regulate safety in areas including workplace health and safety, rail, heavy vehicles, and commercial maritime vessels.

General safety duties under Australian law require companies to ensure the safety of their operations 'so far as is reasonably practicable.' This standard recognises that safety expectations should be high, but not unlimited. The application of the general safety duty to particular scenarios can be developed through advice from regulators and through case law.

'So far as is reasonably practicable' is defined under Workplace Health and Safety Law (and similarly in other regimes) as:

> *that which is, or was at a particular time, reasonably able to be done to ensure health and safety, taking into account and weighing up all relevant matters including:*
>
> *(a) the likelihood of the hazard or the risk concerned occurring*
>
> *(b) the degree of harm that might result from the hazard or the risk*
>
> *(c) what the person concerned knows, or ought reasonably to know, about the hazard or risk, and ways of eliminating or minimising the risk*
>
> *(d) the availability and suitability of ways to eliminate or minimise the risk, and*
>
> *(e) after assessing the extent of the risk and the available ways of eliminating or minimising the risk, the cost associated with available ways of eliminating*

This definition effectively moves operators in the relevant industries from focusing on compliance with specific rules to focus on risks and how they should be mitigated or eliminated.

In June 2020, Australia's transport ministers agreed to key recommendations on in-service safety for automated vehicles in Australia. Ministers agreed Ministers agreed to work towards establishing a single, national approach to regulating automated vehicles when they are on the road [4].

This approach will include:

- a national regulator, and
- a national law,
- supported by a general safety duty.

The agreement to a national law and national regulator is crucial in the Australian context. Australia is a federation, where in-service safety of current vehicles is carried out by state agencies. A national law and regulator ensure a consistent system nationally for companies looking to deploy automated vehicles in Australia. The general safety duty sets clear expectations for industry and provides confidence for consumers.

Under the general safety duty, ADSEs will need to assess the safety risks of their automated driving systems in the context of the operational design domain that the vehicle will operate in. These risks are likely to be different for different operational design domain domains, applications, vehicle types and levels of automation. For example, the risks for a low-speed shuttle operating in pedestrianized areas will be different to those for a heavy vehicle that is capable of automated driving on freeways. The ADSE will then need to work out how to mitigate or eliminate those risks, to meet its general safety duty. The ADSE will also need to continue to monitor for new safety risks or changes to existing risks and respond accordingly. The general safety duty is a dynamic obligation, in recognition of the dynamic environment.

The benefits of a general safety include that it provides significant flexibility to industry – ADSEs can choose how they manage their risks or to change their operational design domain to eliminate certain risks. The general safety duty does not prescribe particular technologies (e.g., cameras v. lidars) or compel certain processes. It also allows companies to change and evolve their operations over time – as long as they continue to operate safely. It avoids the need for prescriptive requirements which can be difficult to comply with and can quickly become out-of-date. Instead of focusing on a myriad of prescriptive requirements, ADSEs can focus their attention on the safety risks. Finally, it ensures that poor operators can be dealt with by regulators [4].

For regulators and the public, the general safety duty ensures a high level of safety that can evolve over time; what's 'reasonably practicable' can evolve as technology evolves. It ensures appropriate penalties for unsafe operations without trying to foresee and specify risks in advance. And it allows regulators to take a pro-active approach – contacting and dealing with ADSEs when a safety risk is identified without having to wait for a crash to occur.

The general safety duty approach changes the paradigm in road transport from prescription (your vehicle must have these features and drive at this speed) to a safety management approach. As a result, it can also change the relationship of the regulator to ADSEs. Current road regulation focuses on punitive actions, such as fines, as we do today for human drivers who breach prescriptive requirements. A general safety duty allows a wider range of responses – including education, warnings, improvement notices and enforceable undertakings – aimed at rectifying the safety issue rather than punishment.

## 5 Conclusions and Further Work

Automated vehicles require new approaches to thinking about regulation. The risks of automated vehicles in-service are different in kind to those of conventional vehicles and require a different response by governments. A principles-based approach, focused on a general safety duty can supports both high levels of safety and innovation.

There are further in-service safety issues to resolve. In particular there is more to work through on in-service issues including:

- how to manage modifications to vehicles in-service, whether by the ADSE or by third parties, including changes to software and hardware
- how to best manage after-market installations of automated driving systems to existing vehicles
- how to develop an appropriate regulator to manage the in-service safety risks.

The National Transport Commission has continued to work with stakeholders to consult on options and propose solutions to these issues.

Automated driving systems create significant risks in-service. These risks must be managed and monitored to ensure that the vehicles can continue to operate safely in a changing external environment. A general safety duty provides the right balance for industry and the public, so that the community can enjoy the benefits of this technology once it is ready for commercial deployment.

## References


1. Society of Automotive Engineers. Taxonomy and definitions for terms related to driving automation systems for on-road motor vehicles (2018)
2. National Transport Commission. Regulatory reforms for automated road vehicles (2016)
3. National Transport Commission. Safety assurance for automated driving systems, decisions regulation impact statement (2018)
4. National Transport Commission. In-service safety for automated vehicles, decisions regulation impact statement (2020)

# Part II: Business Models and Operations

# Local Roadmaps for Autonomous Vehicles: Guidance for High-Impact, Low-Cost Policy Strategies

William Riggs(✉)

University of San Francisco, 2130 Fulton Street, San Francisco, USA
wriggs@usfca.edu

**Abstract.** This chapter focuses on local policy roadmaps. While a significant body of policy has been developed at the national and state levels, very little policy work has identified policy for local government. In this context, this chapter reviews state of technology and policy development and offers potential policy concepts for local government. Suggestions from a workshop are offered and then dialogued in a structured format. This yields ideas for high-impact, low-cost policy solutions that can help cities better prepare for automation.

**Keywords:** Automated vehicles · Local policy · Streets · Design · Management

## 1 Introduction

As automated and autonomous vehicle (AVs) technology and business models continue to accelerate, with level 4 vehicles likely to hit the market by 2025, there is a great deal of progress being made, yet many questions remain. Private car ownership has continued to decline [1] and shared mobility options have continued to accelerate [2, 3]. Companies like Waymo, Cruise and others have logged thousands of hours on roadways in pilot deployments—with advanced sensor systems gathering information about surroundings, sophisticated algorithms that process sensor-based data, and high computational power that controls the vehicle in real-time. This technology offers new business models that have the opportunity to harness the power of platform and place [4].

While these place and partnership-based strategies offer a near term focal point for shared automation, business models will continue to be refined based on preliminary deployments. And as the technologies evolve, the thinking about how AVs could fundamentally shift many aspects of urban travel continues to progress, highlighting the need for nimble policy. Many academics have predicted that there is risk that the rise of private AVs could yield increase VMT and promote urban sprawl [5–7]. Yet with appropriate policy AVs also have the potential to reduce reliance on the personally-owned gasoline-powered car, and have the potential to increase access for historically disadvantaged communities and make cities safer, cleaner, and more sustainable.

In this context city, state, and federal governments are increasingly navigating overlapping and cross-jurisdictional policy landscapes at pace much slower than the rate


G. Meyer and S. Beiker (Eds.): AVS 2020, LNMOB, pp. 49–59, 2022.
https://doi.org/10.1007/978-3-030-80063-5_5

of technological change [8, 9].This provides for a complex policy patchwork spanning municipal jurisdictions. While there has been ample dialogue about automated and autonomous vehicles and policy impacts at a state and federal scale, less discussion has been given to the impacts on local governments and municipal decisions. In this context, this chapter focuses on what cities can do in the next 3–5 years to prepare for autonomous vehicles.

The chapter begins with an overview of autonomous technology. It then focuses on the design and policy landscape both in the US and internationally. Following this it documents policy workshop dialogues and expert panels from at Autonomous Vehicles Symposium (AVS) 2020 highlighting strengths and weaknesses of future AV visions and exploring policy trade-offs. It concludes with a framework for high-impact, low-cost policy steps that local governments can take to prepare for AV deployment, as part of an AV readiness roadmap for local governments.

## 2 Background

### 2.1 The State of Automated Technology

Automated and autonomous vehicle (AVs) technology and business models have continued advance at a fast pace, and most innovators in this space have structured their robotic systems making use of "sense-plan-act" designs [10]. In that sense AVs make use of combination sensors including radar, lidar, infrared, and ultrasonic to sense the surrounding environment. A group of combined sensors can balance one another and cover the weaknesses present in one individual sensor. Robotic and software systems collect data on the environment at which point physical actuators instigate the driving decisions.

In the simplest sense this is how reliable automation is beginning to happen, yet corner cases remain. Some of these relate to how aspects of vehicles design can increase reliability in inclement conditions (for example adverse weather or collision avoidance), particularly with regard to adverse weather conditions. These involve the application of passive (camera) and active (ultrasonic, radar, LIDAR) sensors to dynamically map surroundings and localize its functions in the map as compared to an on-board roadway basemap.

Other corner cases relate directly to aspects of policy that have their roots in local transportation/civil engineering. For example, the reliability and legibility of local signage and roadways can greatly enhance or impede autonomous driving. Similarly, lane controls/access, speeds and other operational domains can help sensors to observe the environment, advanced software to decide and process the path for the vehicle, and actuators to implement the decisions. These corner cases offer the primary opportunity for AV policy.

### 2.2 The State of Policy in the US

In the United States in particular, delayed federal AV legislation has created regulatory and legislative ambiguity. A number of states have advanced a policy framework governing AVs through executive orders, legislation, and regulations. These state AV policy frameworks contemplate everything from testing and deployment to road safety and cybersecurity. As states emerge to fill this policy vacuum left by the federal government, cities are also exploring what role they are to play in an autonomous future. Given the current federal landscape, local and state-level policy efforts are thus exploring what, if any, role they have to play, with few policymakers focused on considering how the built environment may change in a shared, autonomous, and electric future.

### 2.2.1 Federal Policy

While progress is being made, there is still no federal AV legislation in the United States, which has created a regulatory and legal vacuum for states and cities. Legislation like the AV START Act and SELF DRIVE Act would create a national framework for the testing and deployment of automated vehicles that would mirror the existing regulatory precedent for human-driven vehicles. These efforts would clearly delineate the authority of the federal government from that of states and localities, namely over the design, construction, and performance of highly automated vehicles and driving systems. However, these efforts have been critiqued for both substantive and political reasons on the grounds that the proposed frameworks would provide only limited oversight over the development of AVs while preempting states and local governments from regulating the safety of their streets in the face of this innovative technology.

Similarly, the U.S. Department of Transportation (USDOT) has advanced its own guidance, most recently through the December 2019 "Ensuring American Leadership in Automated Vehicle Technologies" report, referred to as Automated Vehicles 4.0 [11]. Their statements offer parallel guidelines to support AV industry maturation, focusing on three high-level principles—protecting users and communities, promoting efficient markets, and facilitating coordinated efforts. Yet it has been similarly critiqued for its focus on voluntary safety consensus, particularly in the wake of high-profile crashes around AV testing. The recommendations are closely tied to the National Highway Traffic Safety Administration (NHTSA) Automated Vehicles 2.0 guidelines which outlined a pathway for safe vehicle operations and provided high-level regulatory guidelines [12, 13].

The US Department of Transportation has taken several other noteworthy steps, including publishing a Request for Comment in 2018 over regulatory barriers to AV testing, a 2019 Advanced Notice of Proposed Rulemaking (ANPRM) on crash protection standards, and a 2020 Notice of Proposed Rulemaking (NPRM) to modernize occupant protection standards for AVs with traditional seating configurations [14].

The policy situation is similar in Europe. The European Commission has sponsored the Horizon 2020 program to advance autonomous development, and the "demonstration of automated driving systems for passenger cars, trucks and urban transport, and on the underlying digital infrastructure to ensure the necessary level of safety, reliability and efficiency." [15] Yet member states of the European Union have developed a patchwork of policy frameworks to allow for testing and preliminary deployment, primarily focused on vehicle standards and safety [16].

### 2.2.2 State and Local Policy

As a result of the current lack of US federal (and EU) guidance, cities and states have stepped in to backfill, advancing their own AV policies, each with unique direction and guidance and subsequent frameworks. This process has created uncertainty and a regulatory patchwork that is unable to keep pace with the rate of technological change.

Meanwhile, of the cities and states that have advanced policies, the absence of federal guidance has led to some policy overlap on issues historically handled at the federal level, such as vehicle safety performance standards. Furthermore, while some non-profit advocacy groups such as the American Planning Association, which outline design and planning principles for AVs in PAS 592, have provided AV-specific design and land use suggestions, there are few policy frameworks specifically for local government [9].

At the local level there is patchwork of policy emerging with many smaller cities acknowledging or mentioning AVs and establishing language in local comprehensive or general plans. At the same time, medium and larger cities have enacted local ordinances and publishing white papers. The variation across geography aligns with how many cities are working to keep up with the pace of technological change but have little window into how it is evolving in the private sector.

Research indicates that, on the whole, regional transportation planning agencies are struggling to keep up with the technology [17, 18]. In 2018, of the top 600 cities in the U.S., 75 (12%) had any dialogue in planning documents on AVs and their impact on urban space. Of these 75, just 29 had a law in the form of an ordinance or general planning principle. The remainder were either white papers or mentions in planning documents.

## 3 A Local Policy Dialogue

Many of these policy frameworks offer the dichotomous scenarios that have been seen in many presentations; extreme visions of heaven or hell involving AVs. These are generalized in Table 1.

A more nuanced approach might look for policy opportunities and roadmaps that guide and control policy development. This reality may be more likely based on survey of participants of the AVS 2020 workshop, highlighting "strengths/weaknesses" and "hopes/fears" about these future visions for automation.

As seen in Table 2, when thematically broken down by their sentiment, the relative perception is one of optimism, alongside the need for policy action. Furthermore, when asked the first thing to come to mind the expert group (N = 60), most jumped to themes related to mobility management strategies; for example managing curbs for deliveries, managing testing and pilots and conducting education.

In the spirit of this theme of managing mobility, there are many policies that can help enhanced accessibility and mobility, while improving sustainability and economic development. They can also help deal with the negative externalities that have been discussed, including potential urban sprawl, congestion, emissions, and long commuting patterns–facilitating wide social, environmental, and economic impacts [6, 19, 20].

One example of these policy synergies is in Chandler, Arizona. Chandler welcomed passenger and testing pilots with the idea of attracting major technology companies to

**Table 1.** Typical scenarios for automated mobility.

| Scenario 1 | Scenario 2 |
|---|---|
| ***Transition towards compact and dense cities with improved life quality*** Shared CAV fleets are integrated within the greater network for public transport that helps cities meet the mobility requirements of residents with nominal private vehicles. Empty parking and road areas are transformed for more productive purposes like green infrastructure, public squares, and affordable housing. Automated vehicles out for delivery function during the night, thereby improving the air quality and reducing congestion. An attractive urban realm promotes business and encourages more cycling and walking, improving public health. On the other hand, shared autonomous fleets of buses in rural areas increase mobility for older people and children, helping fight isolation | ***Transition towards congestion, poor health and greater sprawl*** CAVs get optimized for individual usage. Car journeys become more convenient and productive, with users socializing or resting online and even working remotely during the ride. This promotes a shift distant from public transportation as well as urban living to affordable suburbs. People make more frequent and longer journeys today to access services and jobs. This results in large sections of the road network becoming congested quickly. Roads are expanded to cater to the rising demand that eases urban sprawl through increased accessibility to affordable land on city and town outskirts. Roads become separated from cyclists and pedestrians to help CAVs travel without disruption. Long hours of inactive commuting mean less time for people to socialize, prepare food, and exercise, leading to a decline in public health |

**Table 2.** Survey participant sentiments (N = 60).

| Positive | Negative |
|---|---|
| ***Sentiment for policy experts with regard to AVs*** | ***Sentiment for policy experts with regard to AVs*** |
| Safe/bike & pedestrian Safety (x4) | Increased VMT (x3) |
| Shared (x3) | Unintended consequences (x3) |
| Exciting (x2) | Difficult |
| Electric (x3) | Not. Quite. Yet |
| Opportunity/welcome (x2) | Complex |
| Access/increased Mobility (x2) | Contentious |
| Freedom (x2) | |
| Behavior change | |
| Slow | |
| Overdue | |
| Sensing | |
| Sustainable | |
| Multitasking | |
| Land Use | |
| Equity | |
| Utopia | |
| ***Management strategies*** | |
| Logistics/contactless delivery (x3) | |
| Curb/parking management (x2) | |
| Testing (x2) | |
| Education | |

situate their research and development within city premises and establish AV pilots. Likewise, Las Vegas, Nevada and San Jose, California have created innovation policies that offer proving grounds for rising technologies. They have worked to eliminate barriers for on-road testing. San Jose specifically has included thinking about AVs in their transport system—bringing future deployments in line with the existing goals of the municipality. As the third largest city in California, it adopted this plan to better understand how the AV companies might operate with the city and create chances for alignment with transportation goals.

These thinking about mobility management also extends to streets and curbs – particularly since in an idea world AVs should ease the pick-up and drop-off experience—making it safer and more efficient [21]. Seniors who cannot drive traditional vehicles safely, can be mobile due to these autonomous vehicles, and everything from the curbs to sidewalks to the streets can be changed to facilitate their ingress and egress [22–25]. This impacts a number of factors related to street design which local government can facilitate by focusing on things like:

- Infrastructure
- Curb management
- Public space and positive sum roadway designs

### 3.1 Infrastructure

The most vital infrastructure required for automated vehicles are well maintained streets. Properly marked lines, well-defined curbs, smooth pavements, and proper signs offer easy readability for AVs. This involves local prioritization of funding to ensure that this kind of maintenance can be conducted but also looking forward to what infrastructure can become. For example, 3M's Connected Roads program, are exploring dedicated paint only visible in the infrared that can alert AV systems via RFID codes to upcoming construction zones [26].

Local infrastructure policy for AVs should also includes energy. Cities can broaden access to EVs as urban mobility evolves. In states such as California, Zero-Emission Vehicle mandates have helped accelerate both industry maturation and technology development, and are poised to significantly decarbonize transportation. This offers an opportunity to consider local policy incentives to support vehicle financing for lower-income residents, as well as "make-ready" requirements for new construction projects.

### 3.2 Curb Management

With a significant number of residents making a shift to Transportation-as-a-Service from ownership, the demand for drop-off and pickup on the streets will continue to grow. Spaces can be priced for access. Curb spaces can be prioritized for high occupancy vehicles, disabled people and buses. Cities should designate dedicated pick-up and drop-off locations that prioritize shared and/or electric transportation modes, as well as non-automotive travel.

Curbs should be "high productivity" in that they should maximize and promote available technology to the greatest extent possible. Possible solutions could include utilization of sensors, digital signage, and integrated navigation apps to communicate real-time information for planners, developers, and engineers.

### 3.3 Public Space and Positive Sum Roadway Design

Positive sum roadway design that captures space for public uses and non-automotive travel is essential for urban travel, and consistent with the move in many cities to consider the "15 min City." This can include pedestrian zones, which cities around the world have begun to explore—for example Barcelona, Madrid, and Oslo. Similarly, positive sum design is aligned with Vision Zero initiatives—which emphasizes the importance of reassigning roadway space in a way that moves beyond "relocation" of uses or conversion but that creates flexible roadways that can be used for public space and multimodal travel.

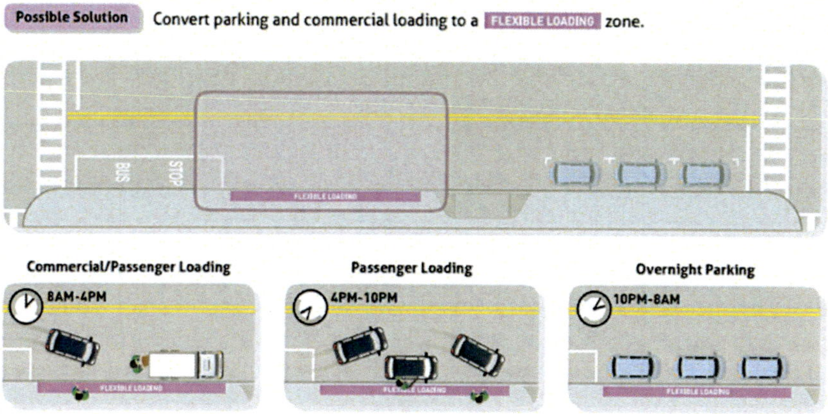


**Fig. 1.** A flexible possible solution for AV streets by Fehr & Peers.

As shown in Fig. 1, AVs come with the opportunity to free a considerable amount of roadway lanes and make them more flexible; reducing widths of certain travel lanes and making on-street parking more agile. This space could conceivably become more than a location just for transportation as urban space evolves—for example it could include play areas, parks/open space or future buildings [27, 28].

## 4 Roadmaps for the Future

Developing local policy for AVs involves building on existing strategies and helping cities develop plans to transform alongside AVs. A primary way that planners and engineers can explore this transformation is by thinking of the street as an asset and thinking about a continuum of low-cost, short-term activities practices that can form a roadmap for local AV policy.

As shown in Fig. 2, at the most basic level this involves positive sum roadway designs and incorporation with existing policy. This means doing parking and curb management, building infrastructure for EVs and maintaining road markings; pursuing smart/shared streets that support cycling and walking, and thinking about the future of smart infrastructure (while at the same time not over-thinking this infrastructure).

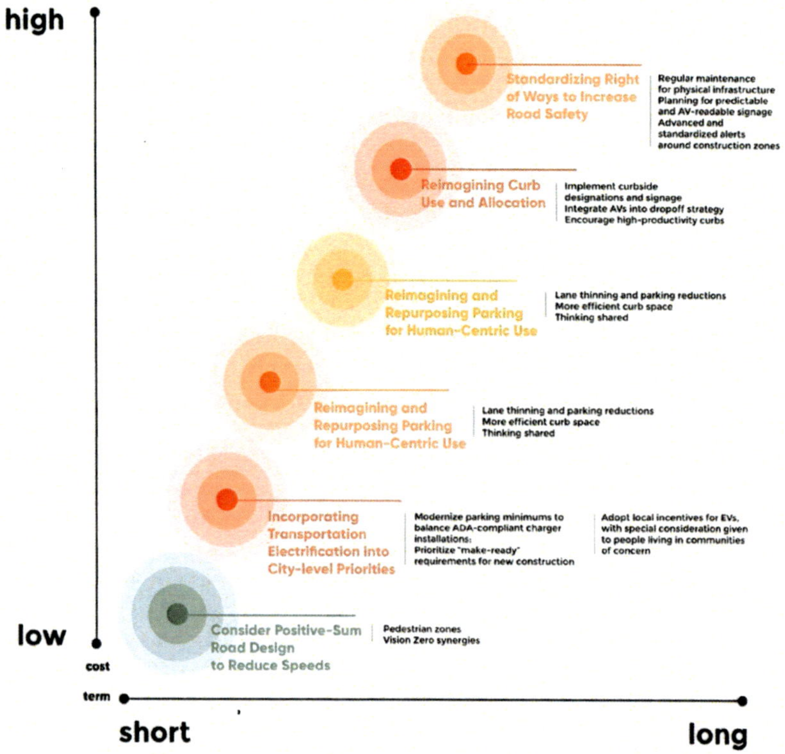


**Fig. 2.** A continuum of policy from low-cost, short term to high-cost, long-term [29].

In this aspect local governments can take time to focus on core infrastructure. They can: have clear signs, lines and good pavement; consider multiple roadway users and the implications for bikes/pedestrians; prioritize share vehicles on more heavily trafficked arterials and collectors; and think about mobility corridors that might include AVs as a way to increase mobility and access. These mobility corridors could provide the framework for AV corridors in cities; offering a way to increase the number of deployments within the patchwork policy framework spanning local governments.

Over the long-term AV transportation also promises local governments opportunities for higher-cost transformations. There is potential to harness streets as a real estate asset that requires maintenance and management, and to shape streets based on the lessons from the COVID-19 pandemic to be more bicycle and pedestrian friendly [30]. There are opportunities for local governments to consider directionality and the future of one-way streets [31, 32]. There are new finance mechanisms for cities with reduction of revenue from traditional sources (parking, fuel tax, etc.) [33, 34].

These aspects of automation at the local level illustrate that cities have a unique window to define their own role and chart an equitable, green, accessible, and inclusive transportation vision that is fundamentally rooted in the policy areas where they have expertise. Local governments can lead this charge with proactive and constructive dialogues that best capture the impacts of new mobility on their communities.

## References

1. Naughton, W.: This is What Peak Car Looks Like. Bloomberg.com, 28 February 2019. https://www.bloomberg.com/news/features/2019-02-28/this-is-what-peak-car-looks-like. Accessed 01 Feb 2020
2. Shaheen, S.: Shared mobility: the potential of ride hailing and pooling. In: Sperling, D. (ed.) Three Revolutions, pp. 55–76. Island Press/Center for Resource Economics, Washington, DC (2018). https://doi.org/10.5822/978-1-61091-906-7_3
3. Yakovlev, A., Otto, P.: The Future of Mobility - Shared Mobility. Ipsos (2018). https://www.ipsos.com/en-vn/future-mobility-shared-mobility. Accessed 01 Feb 2020
4. Riggs, W., Beiker, S.A.: Business models for shared and autonomous mobility. In: Meyer, G., Beiker, S. (eds.) AVS 2019. LNM, pp. 33–48. Springer, Cham (2020). https://doi.org/10.1007/978-3-030-52840-9_4
5. Riggs, W.: Disruptive Transport: Driverless Cars, Transport Innovation and the Sustainable City of Tomorrow. Routledge, London (2019)
6. Riggs, W., Larco, N., Tierney, G., Ruhl, M., Karlin-Resnick, J., Rodier, C.: Autonomous vehicles and the built environment: exploring the impacts on different urban contexts. In: Meyer, G., Beiker, S. (eds.) AUVSI 2018. LNM, pp. 221–232. Springer, Cham (2019). https://doi.org/10.1007/978-3-319-94896-6_19
7. Airbib, J., Seba, T.: Rethinking Transportation 2020–2030: The Disruption of Transportation and the Collapse of the Internal-Combustion Vehicle and Oil Industries. RethinkX (2017)
8. Chatman, D.G., Moran, M.: Autonomous Vehicles in the United States: Understanding Why and How Cities and Regions Are Responding, August 2019. https://escholarship.org/uc/item/29n5w2jk. Accessed 24 Sep 2019
9. Crute, J., Riggs, W., Chapin, T., Stevens, L.: Planning for Autonomous Mobility. American Planning Association, Washington D.C. (2018). PAS 592. https://www.planning.org/publications/report/9157605/
10. Anderson, J.M., Kalra, N., Stanley, K.D., Sorensen, P., Samaras, C., Oluwatola, O.A.: Autonomous Vehicle Technology: A Guide for Policymakers. Rand Corporation, Santa Monica, CA (2014)
11. USDOT: AV 4.0 (2019). https://www.transportation.gov/policy-initiatives/automated-vehicles/av-40. Accessed 19 Jan 2021
12. NHTSA: USDOT Automated Vehicles 2.0 Activities (2017). https://www.transportation.gov/av/2.0. Accessed 19 Jan 2021
13. NHTSA: Federal Automated Vehicles Policy: Accelerating the Next Revolution in Roadway Safety. US Department of Transportation (2016). https://www.transportation.gov/sites/dot.gov/files/docs/AV%20policy%20guidance%20PDF.pdf
14. Daniels, P.: Statement on U.S. DOT's 4.0 Autonomous Vehicle (AV) Policy, 08 January 2020. https://saferoads.org/2020/01/08/statement-on-u-s-dots-4-0-autonomous-vehicle-av-policy/. Accessed 19 Jan 2021
15. European Commission: Automated road transport. Innovation and Networks Executive Agency - European Commission, 02 December 2015. https://ec.europa.eu/inea/en/horizon-2020/automated-road-transport. Accessed 16 May 2021

16. AutoVista Group: The state of autonomous legislation in Europe. Autovista Group (2019). https://autovistagroup.com/news-and-insights/state-autonomous-legislation-europe. Accessed 16 May 2021
17. Guerra, E.: Planning for cars that drive themselves: metropolitan planning organizations, regional transportation plans, and autonomous vehicles. J. Plan. Educ. Res. **36**(2), 210–224 (2015). https://doi.org/10.1177/0739456X15613591
18. Riggs, W.: Benchmarking automated and autonomous vehicle policies in the United States. In: Meyer, G., Beiker, S. (eds.) AVS 2019. LNM, pp. 144–160. Springer, Cham (2020). https://doi.org/10.1007/978-3-030-52840-9_14
19. van Arem, B., et al.: Building automation into urban and metropolitan mobility planning. In: Meyer, G., Beiker, S. (eds.) AVS 2018. LNM, pp. 123–136. Springer, Cham (2019). https://doi.org/10.1007/978-3-030-22933-7_13
20. Guerra, E.: When autonomous cars take to the road. Planning **81**(5) (2015). https://trid.trb.org/view.aspx?id=1358466. Accessed 13 Apr 2016
21. Appleyard, B., Riggs, W.: 'Doing the right things' before 'doing things right': a conceptual transportation/land use framework for livability, sustainability, and equity in the era of autonomous vehicles. Presented at the Transportation Research Board 97th Annual Meeting Transportation Research Board, Washington, D.C. (2018). https://trid.trb.org/view/1496858. Accessed 01 Aug 2018
22. Schlossberg, M., Riggs, W.W., Millard-Ball, A., Shay, E.: Rethinking the street in an era of driverless cars. UrbanismNext (2018). https://urbanismnext.uoregon.edu/files/2018/01/Rethinking_Streets_AVs_012618-27hcyr6.pdf
23. Leistner, D.L., Steiner, R.L.: Uber for seniors?: exploring transportation options for the future. Transp. Res. Rec. **2660**(1), 22–29 (2017). https://doi.org/10.3141/2660-04
24. Riggs, W., Appleyard, B., Johnson, M.: A design framework for livable streets in the era of autonomous vehicles. Urban Plan. Transp. Res. **8**(1), 125–137 (2020). https://doi.org/10.1080/21650020.2020.1749123
25. Riggs, W., Appleyard, B., Johnson, M.: A design framework for livable streets in the era of autonomous vehicles. Presented at the Transportation Research Board, Washington, D.C. (2019)
26. Hyatt, K.: 3M Connected roads aim to make life easier for autonomous vehicles. Roadshow (2018). https://www.cnet.com/roadshow/news/3m-connected-roads-aim-to-make-life-easier-for-autonomous-vehicles/. Accessed 21 Jan 2021
27. Riggs, W.W.: Modeling Future Street Options in an AV Future Using Restreet (2017)
28. Riggs, W.: Technology, civic engagement and street science: hacking the future of participatory street design in the era of self-driving cars. In: Proceedings of the 19th Annual International Conference on Digital Government Research: Governance in the Data Age, Delft, Netherlands, pp. 4:1–4:6 (2018). https://doi.org/10.1145/3209281.3209383
29. Riggs, W.: A Policy Framework for the Future of Automated Mobility: The Need for Local Government Policy. Mineta Transportation Institute, San Jose, Project 2055, October 2020. https://transweb.sjsu.edu/research/2055-Policy-Framework-Future-Automated-Mobility. Accessed 21 Jan 2021
30. Riggs, W.: Telework and sustainable travel during the COVID-19 era. Social Science Research Network, Rochester, NY, June 2020. SSRN Scholarly Paper ID 3638885. https://papers.ssrn.com/abstract=3638885. Accessed 31 Jul 2020
31. Riggs, W., Gilderbloom, J.: Two-way street conversion: evidence of increased livability in Louisville. J. Plan. Educ. Res. **36**(1), 105–118 (2016). https://doi.org/10.1177/0739456X15593147

32. Riggs, W.: Reduced perception of safety for cyclists on multi-lane, one-way and two-way streets: opportunities for behavioral economics and design. Social Science Research Network, Rochester, NY, August 2017. SSRN Scholarly Paper ID 3011680. https://papers.ssrn.com/abstract=3011680. Accessed 03 Oct 2017
33. Riggs, W., Appleyard, B.: The economic impact of one to two-way street conversions: advancing a context sensitive framework. Transp. Res. **18**, 19 (2016)
34. Clark, B., Larco, N.: The Impacts of Autonomous Vehicles and E-commerce on Local Government Budgeting and Finance. UrbanismNext (2018). https://urbanismnext.uoregon.edu/files/2017/07/Impacts-of-AV-Ecommerce-on-Local-Govt-Budget-and-Finance-SCI-08-2017-2n8wgfg.pdf

# Artificial Intelligence for Automated Vehicle Control and Traffic Operations: Challenges and Opportunities

David A. Abbink[1], Peng Hao[2], Jorge Laval[3], Shai Shalev-Shwartz[4], Cathy Wu[5], Terry Yang[6], Samer Hamdar[7](✉), Danjue Chen[8], Yuanchang Xie[9], Xiaopeng Li[10], and Mohaiminul Haque[7]

[1] Section Human-Robot Interaction, Department of Cognitive Robotics, Faculty 3mE, Delft University of Technology, Room: 34-F-2-120, Mekelweg 2, 2628, CD Delft, The Netherlands
D.A.Abbink@tudelft.nl

[2] University of California Riverside, Riverside, USA
peng.hao@ucr.edu

[3] Georgia Institute of Technology, SEB 224, 788 Atlantic Drive NW, 30332 Atlanta, GA, Georgia
jorge.laval@ce.gatech.edu

[4] Mobileye and The Hebrew University of Jerusalem, Rothberg Building, Room B428, Givat Ram, 91904 Jerusalem, Israel
shais@cs.huji.ac.il

[5] Massachusetts Institute of Technology, 77 Massachusetts Avenue, Cambridge, MA 02139-4307, USA
cathywu@mit.edu

[6] The University of Utah, Floyd and Jeri Meldrum Civil Engineering, 110 Central Campus Drive, UT 84112 Salt Lake City, USA
x.yang@utah.edu

[7] George Washington University, 800 22nd Street NW, Washington, DC 20052, USA
{hamdar,mohaiminul}@gwu.edu

[8] University of Massachusetts Lowell, 1 University Avenue, PA 108, MA 01854 Lowell, USA
danjue_chen@uml.edu

[9] University of Massachusetts Lowell, 1 University Avenue, MA 01854 Lowell, USA
Yuanchang_Xie@uml.edu

[10] University of South Florida, 4202 E. Fowler Avenue, ENG 207, Tampa, FL 33620-5350, USA
xiaopengli@usf.edu

**Abstract.** This chapter summarizes the presentations of speakers addressing such issues during the Automated Vehicles Symposium 2020 (AVS20) held virtually on July 27–30, 2020. These speakers participated in the break-out session titled "Artificial Intelligence for Automated Vehicle Control and Traffic Operations: Challenges and Opportunities". The corresponding discussion and recommendations are presented in terms of the lessons learned and the future research directions to be adopted to benefit from AI in order to develop safer and more efficient connected and automated vehicles (CAV). This session was organized by the Transportation Research Board (TRB) Committee on Traffic Flow Theory and Characteristics (ACP50) and the TRB Committee on Artificial Intelligence and Advanced Computing Applications (AED50).


G. Meyer and S. Beiker (Eds.): AVS 2020, LNMOB, pp. 60–72, 2022.
https://doi.org/10.1007/978-3-030-80063-5_6

**Keywords:** Traffic flow modeling · Traffic operations · Control · Automated vehicles · Artificial intelligence

# 1 Introduction

Artificial Intelligence (AI) models are being utilized extensively in different scientific and engineering domains for both analysis and predictive purposes. In particular, AI models are leveraged for processing sensor data, controlling automated vehicles (AV) and operating traffic control devices. However, there are still many challenges in those AI applications, including how to choose, build, and train AI models to avoid issues such as overfitting; translate AI models trained on synthetic (e.g., simulated) data to real-world applications; and teach AI controlled AVs how to collaborate (instead of solely maximizing their own benefits) with each other, human-driven vehicles, and traffic control devices at both local and network levels so that the overall transportation system's safety and mobility are maximized.

Among the scientific and engineering domains using AI models, automotive makers are adopting AI techniques in order to automate the movement of driver-less cars thus creating safer and more reliable automated vehicles (AV). On the other hand, traffic engineers are adopting AI to predict congestion and collision formation on our roadway networks offering real-time information for users to make better travel decisions. However, simply adopting AI instead of standard traffic flow models may lead to the lack of understanding of physical processes and dynamics leading to poor roadway performance and may produce false predictions in traffic states especially given the need of extensive data for AI training, calibration and validation purposes. Accordingly, the suitable use of AI in traffic operations and AV models requires studying two dimensions: 1) the type of AI models being adopted and their corresponding characteristics; and 2) the gap between the data available for transportation professionals and the data needed to train AI models.

Towards studying the AI model characteristics and the corresponding data needs, the Transportation Research Board (TRB) Committee on Traffic Flow Theory and Characteristics (ACP50) and the TRB Committee on Artificial Intelligence and Advanced Computing Applications (AED50) organized a breakout session at the Automated Vehicles Symposium 2020 (AVS20) - held virtually on July 27–30, 2020. The breakout session titled "Artificial Intelligence for Automated Vehicle Control and Traffic Operations: Challenges and Opportunities" brought together six scholars from academia and the industry. These scholars presented their latest work in AI as related to the traffic engineering and AV field. Following the presentations, a panel consisting of five of the invited speakers had extensive discussion with the audience. This chapter summarizes these presentations while identifying the key challenges in adopting AI for traffic and AV modeling and the corresponding efforts made to adapt data for training and calibration purposes. In particular, the objectives of the session are to:

- Identify the opportunities and challenges associated with AI applications in AV control and traffic operations

- Propose solutions for addressing the challenges
- Identify innovative applications made possible by AI and AV
- Explore how AI can enable collaborative behaviors and their impacts on transportation

Towards realizing the aforementioned objectives, the remaining sections of this chapter are organized as follows: Sect. 2 presents a summary of the 6 invited talks and Sect. 3 introduces the key results from the panel discussion.

## 2 Research on Utilizing Artificial Intelligence (AI) for Traffic Operations and Automated Vehicle (AV) Modeling

This section presents a summary of the six invited talks, which addressed the research challenges, opportunities and existing efforts in adapting AI to design better Automated Vehicles (AV) and to capture their impact on traffic flow. The summary includes the motivation and contributions associated with the presented research, the main conclusions, and future research directions.

### 2.1 The Value of Good Old-Fashioned Parametric Models for AV Control[1]

Adopting artificial intelligence (AI) and self-learning algorithms (SLAs) have had a significant modelling impact on the development of vehicle automation systems. Despite having great potential, such adoption has some limitations. AI and SLAs are usually data hungry and do not behave well in new situations without proper training. This can lead to major safety issues when deploying Automated Vehicles (AV). As long as no extensive data repositories are provided to AV developers covering a wide range of traffic conditions, there will be misalignments between vehicle dynamics/movements controlled by SLAs and human driving behavioural adaptations. Such misalignments might be caused by (1) AI anomalies leading to unpredictable "harmful" movements; and (2) interactions between human and AV systems/interfaces due to lack of training and communication. As parametric models tend to capture the underlying human driving behaviour with specific modelling and theoretical constructs, they can play a role to mitigate such misalignments. A three-pronged approach (Melman et al. 2020) might be suggested to address this problem: *i)* mitigating misalignments by modelling realistic driving behaviour; *ii)* including parameterized models of driver behaviour adaptation into interaction design; and *iii)* offering human-centred interaction design.

i) *Modelling realistic driving behaviour:* Driver behavioural understanding is essential to modelling realistic driving behaviour. Recently a quantification of Gibson's safe field of travel has been proposed (Kolekar et al. 2020a) as the underlying principle for a generalizable driver model. In order to compute a perceived risk, this theory evaluates the consequences/utilities of events occurring in the driving scene and the driver's subjective belief related to the probability of an event to occur. Combined with an assigned weight, the model quantifies the perceived risk and is able to describe and predict different naturalistic driving behaviours in various traffic scenarios (Kolekar et al., 2020b).

[1] By David A. Abbink, Delft University of Technology, Netherlands.

One major benefit of this model is that it can perform well in unobserved situations with no readily available data. Accordingly, this modelling approach can contribute to mitigating one of the major problems associated with AI based autonomy.

ii) *Including models of driver behaviour adaptation into interaction design:* Drivers might adopt undesirable or risk seeking behaviour when using SAV (Semi-Automated Vehicles) or AV. Melman et al. (2017) showed in their study that drivers trend to drive faster with the haptic lane keeping assistance system. To mitigate such type of emergent behaviour, in-depth human factor and behavioural adaptation studies are necessary.

iii) *Offering human-centred interaction design:* researchers may offer breakthroughs associated with AV development and deployment; however, even when the proposed AV systems do offer a perfect safety record and a significant efficiency improvement, there will always be need for human-automation interaction. Human-automation interaction can be categorized into two categories: traded control and shared control. In traded control, at any specific time during the driving event, either the algorithm or the human controls the vehicle. This approach is comparably easy to implement and computationally less complex. In shared control, human and algorithm can both control the vehicle at any given time. Abbink et al. (2012) demonstrated one such system in which torques on the steering wheel is used for the interaction between the human driver and the algorithm. This torque is used to inform human driver about the disagreement between the trajectories produced by the algorithm and human. In the aforementioned study, the time to lane crossing (TLC) is measured for a human controlled vehicle, a shared controlled vehicle and a traded controlled (automated) vehicle through a simulator environment and it is found that the shared controlled vehicle always performs better. Moreover, in the case of automated system failure, the shared controlled system performs better because it takes less time for humans to take over the control of the vehicle and to react to the situation at hand (if compared to the traded controlled system).

In conclusion, the "old-fashioned" parametric models can play a critical role in solving misalignments between self-driving algorithms and humans, while adapting the human decision-making process to the automation technology and increasing AV safety.

## 2.2 Deep Learning Based Eco-driving for Connected and Automated Vehicles[2]

Human and freight transportation is one of the most energy-consuming sectors in the United States (US). According to a survey by the Energy Information Administration (EIA), about 28% of the total US energy consumption was associated with the transportation sector in 2019. To develop a more energy efficient and sustainable transportation system, the Connected and Automated Vehicle (CAV) technology emerges as one of the transformative solution approaches to such an environmental problem. Connected eco-driving refers to the connected and automated driving system that seeks to minimize the expected total vehicle energy consumption by taking optimal and valid actions. This system also takes into consideration other factors such as maximum travel time, fuel and battery cost. Current research in energy efficient vehicles modelling can be divided into three categories based on the methods used: rule-based models, optimization-based models, and deep-learning models. Usually, the rule-based models are simple to implement,

[2] By Peng Hao, University of California Riverside, U.S.A.

computationally efficient and suitable for real-time use. However, they are designed based on the assumptions and the experience of the researchers and cannot guarantee that the solution is the best (optimal) strategy among all possible alternative strategies. The optimization-based models can define objectives and search and find the best (local or global) alternative strategy. However, these models are computationally complex and most of them are not suitable for real-time and real-world implementation. Moreover, optimization-based models often underestimate the impact of exogenous aspects of the driving environment (i.e., not considered in the formulation either in the objective function or in the constraints).

To address the limitations of the optimization-based models, a graph based modular hybrid model is introduced by Hao et al. (2015) which uses graph theory models and learning-based modules. This model adopts a different approach (including machine learning algorithms) for each of the modules to archive the optimal speed and trajectory plan for energy efficient driving. It includes long short-term memory (LSTM) based signal timing prediction, radial basis function neural-network-based speed forecasting, machine learning based trajectory planning algorithms (MLTPA), etc. for real-time and effective execution of the model.

Deep learning-based modules utilize different deep learning algorithms for connected eco-driving. There are three different logical tasks associated with such modules: *i)* energy efficiency, *ii)* interaction with other traffic units, and *iii)* interaction with infrastructure. One of the challenges encountered is the implementation of these three different logical tasks in the same deep-learning construct. Hao et al. (2020) introduce a hybrid reinforced learning-based approach for eco-driving at a signalized intersection. Markov Decision Process (MDP) is used in their study to address the challenge of implementing three logical tasks in one problem. MDP is a mathematical framework that can be used to model decision making based on the interaction between the learning agent and its environment.

Dueling Deep Q Network (DDQN) is found to be the best among all neural networks studied by Hao et al. (2020). The agent vehicle has on-board sensors for knowing its current state as well as the surrounding traffic environment. It receives V2I (vehicle to infrastructure) information using Dedicated Short-Range Communication (DSRC) system or 5G cellular data. An on-board computer equipped with the decision manager algorithm calculates the long short-term reward of an action (to maximize an objective function over the whole trip instead of the immediate next few steps). The neural network model proposed by Hao et al. (2020) has two main components: a hidden feature extraction component and a policy network which is based on the DDQN architecture. Unity 3D is used to create a virtual reality environment for testing the proposed system. Three types of vehicle agents (governed by three models) are implemented in the virtual reality simulation: an intelligent driver model vehicle, a fast-speed model vehicle (always seeking to maximize the speed) and the eco-driving model vehicle based on DDQN. Results from the study show that the DDQN deep learning model vehicle performs better in terms of energy efficiency. The vehicle also has a smoother acceleration and deceleration pattern and an improved lane-changing performance if compared to the other two vehicles.

In conclusion, the deep learning-based model shows great potential in developing eco-driving strategies for CAV. CAV in eco-driving mode can significantly reduce energy consumption in the transportation sector and help move towards a more sustainable transportation system.

### 2.3 Machine Learning Methods – Beware: There is Nothing to Learn About Congested Urban Networks[3]

Congested urban networks have long been considered to behave chaotically and to be very unpredictable. This apparent complexity has led to the development of numerous signal control algorithms, mathematical programs and learning-based control methods to optimize network performance. Most of the research shows some operational improvements but they mostly correspond to light traffic conditions or very specific small networks. The recent empirical verification of the existence of a network-level Macroscopic Fundamental Diagram (MFD) suggests a different result when studying congested networks. The network MFD is a way of describing the traffic flow of urban networks at an aggregate level which is used for displaying network simulation output in a concise way. Though the turning probability at intersections is also a key variable that significantly affects the MFD, it is not well understood in the research. Moreover, there is a gap in the deep reinforcement learning (DRL) literature associated with the analysis of the different aspects of large traffic flow networks that influence the performance of DRL methods. It is not clear if and how network congestion levels affect the learning process, nor if other machine learning methods are effective, nor if current findings also apply to large networks.

Laval and Zhou (2020) provide additional evidence for the congested network property of the MFD and analyze how these properties affect the performance of machine learning methods applied to signal control. The traffic flow model used in this study is a cellular automaton (CA) implementation of the kinematic wave model with a triangular flow-density fundamental diagram, which is the simplest model able to predict the main features of traffic flow. A grid network of bidirectional streets with one lane per direction and with a traffic light at all intersections is used as the simulation network. To attain spatial homogeneity, the network is defined on a torus where each street can be thought of as a ring road where all intersections have 4 incoming and 4 outgoing approaches. Vehicle routing is set to random. A driver reaching the stop line, will choose to turn with probability $p$ or keep going straight with probability $1 - p$. Traffic signals in the simulation operate under the simplest possible setting with only red and green phases (no lost time, all-red, yellow nor turning phases). All the control policies considered are incremental in the sense that decisions are taken every $g$ time steps, which can be interpreted as a minimum green time: After the completion of each green time of length $g$, the controller decides whether to prolong the current phase or to switch light colors.

The baseline experiment shows that urban networks are more predictable than previously thought with respect to signal control and the network throughput is independent of traffic signal control even for inhomogeneous networks. To analyze the performance of AI methods, three methods are used for training the signal control policy: random

[3] By Jorge Laval, Georgia Institute of Technology, U.S.A.

search, supervised learning and DRL. The random search method shows that all policies, no matter how inefficient, are optimal when the density exceeds approximately 75% indicating that the network throughput is independent of traffic signal control. Supervised learning training the policy with only two examples yields a near-optimal policy. The simulations indicate that DRL policies are only competitive and lose their ability to learn a sensible policy as the training density increases. Such a finding also indicates that the more the congestion is, the less the policy affects intersection throughput.

The main takeaway from the study is that, on congested urban networks, intersection throughput tends to be independent of signal control. It can be conjectured that this prevents DRL methods from finding sensible policies under congested conditions. In other words, all the DRL methods proposed in the literature to date may be unable to learn sensible policies and may deteriorate as soon as congestion appears on the network.

## 2.4 On the Challenges of Building a Camera-Only, Complete, Self-driving System[4]

Current technologies have either sophisticated technologies with low accuracy requirement or simple technologies with high accuracy requirements. Automated Vehicle (AV) is both a sophisticated technology and requires extremely high accuracy detection. For human drivers, the accidents that involve injuries and/or fatalities occur approximately every $10^4$ h and $10^6$ h respectively. Accordingly, to improve safety, AV should have a mean time between failures (MTBF) to be at least $10^7$ h. The challenge is then to achieve such a high accuracy AV system and to validate such a system with appropriate data. The AV system has three phases: sensing, planning and acting. In the sensing phase, the AV system, with the help of different sensors, builds a three-dimension (3D) environment surrounding the corresponding vehicular space. In the planning phase, it analyzes such environment and finds the optimal driving strategy. In the acting phase, it executes the actions identified in the planning phase.

AVs need very high accuracy sensing for the required large MTBF (needed for safety considerations). In order to tackle such sensing challenge, the concept of redundancy is used. The approach is to build two fully independent subsystems, one with only cameras and another with radars and lidar. These two independent sub-systems should aim to reach a MTBF of $10^4$ h each. If the two sub-systems are truly independent, even in the worst-case scenario (with a MTBF of $10^{3.5}$ h), the sensing technology can achieve the safety standards of $10^7$ h MTBF. The objective becomes building an only camera-based subsystem that can reach a MTBF of $10^4$ h. The main challenges in camera-based sensing come from the fact that cameras are inherently two-dimension (2D) systems and yet we need a 3D understanding of the surrounding environment with high accuracy even in edge cases (where visibility is very low). There are several methods to produce 3D data from 2D camera like prediction of object dimension, visual lidar (VIDAR)/structure from Motion, etc. VIDAR uses deep learning to generate a 3D model of the environment from the camera feed. Another approach is to project the 3D map information into a 2D image plane and then use the 2D data from the camera for the planning phase. The

[4] By Shai Shalev-Shwartz, Mobileye.

redundancy approach uses multiple detection methods like VIDAR, scene segmentation 3DVD, etc. and multiple measurement techniques.

The planning phase involves decision making to find optimum actions to avoid accidents. However, to find actions avoiding accidents at all cost may not be an ideal solution for this phase. There should be a balance between being a perfect driver and being a driver who blends in. This also raises some ethical questions. The usefulness/safety tradeoff adopted by humans requires a sense of caution. The duty of care (Tort Law) states that a legal obligation is imposed to an individual rewiring adherence to a standard of reasonable care while performing any act that could foreseeably harm others. Human has common sense to interpret such standard/law. The challenge is to interpret the law for the AV systems. Rigorous mathematical modeling is required to formalize an interpretation of any law which is applicable to AVs. The resulting models should be sound, useful and applicable to machines. Soundness implies that the interpretation by the model should comply with the common sense of human driving. The interpretation should lead to zero accidents in a utopic world. Usefulness ensures that AVs don't block traffic being over-cautious and non-agile. Efficiency of AV models should be verified for machine applicability. This is not trivial due to potential of butterfly effect. Mobileye proposes a system called Responsibility Sensitive Safety (RSS) model, which is a mathematical model to formalize a commonsense interpretation of the duty of care. RSS should provide mathematical guarantees for AV to never cause an accident, to be relevant to human drivers and to be efficiently verifiable.

In summary, to tackle the challenges of achieving AV safety standards, Mobileye focuses on *1)* redundancy for 3D sensing, and *2)* formal safety modeling during planning (i.e., planning phase) while considering human judgement considerations.

## 2.5 Mixed Autonomy Traffic: A Reinforcement Learning Perspective[5]

We imagine that the future of the transportation sector relies on fully automated and highly efficient transportation systems. It is predicted that by the year 2050 we will achieve full autonomy for surface vehicles. We have multiple billion-dollar corporations racing for the creation of the first fully automated vehicle (AV) and they are improving year after year towards reaching such a goal. There are many tools available for analyzing a single AV with full autonomy while adopting deterministic models with no uncertainty/error. The operation of a single AV depends however on other vehicles in the system and there is a need for additional studies on its impact on the whole mixed (i.e., with the existence of both automated and human driven vehicles) system. In other words, there exists a significant challenge represented by the understanding of and modeling the mixed autonomy state of the transportation system. This challenge is due to the existence of many sources of uncertainty including partial observation, limited communication, data collection challenges, etc.

Mixed autonomy can take different forms like advanced driver assistance systems. The impact of such mixed autonomy on safety, reliability efficiency, fairness should be analyzed. Understanding the impact of mixed autonomy on broader societal system is also necessary. All of these issues require more analysis tools. To analyze the problem

[5] By Cathy Wu, Massachusetts Institute of Technology, U.S.A.

at hand, deep reinforcement learning (DRL) is used. In this modeling approach, agents are the vehicles that are automated and everything else is considered as the exogenous environment. The agents will make decisions such as when to accelerate or decelerate according to a learned policy in order to maximize reward. The average velocity, energy consumption, travel time, safety and comfort should be considered in the global reward function. Ultimately the goal is to study large urban networks where a fraction of the vehicles is automated. Wu et al. (2017) explores the potential of DRL methods when training the algorithm from scratch. This study designs a representative set of scenarios that exhibits a variety of different traffic phenomena including intersection, bottleneck and on/off ramp scenarios. By using DRL with 5–10% of AVs, the simulations show from 30% to 142% increase in average velocity across the scenarios. Some of the learned policies match the performance and behavior of the control strategies devised by experts over the years. These findings provide validation of the methodology to analyze the impact of mixed autonomy in urban environments. A critical challenge of analyzing these systems in large-scale contexts comes from the fact that no two cities' traffic networks are the same; even the traffic network in a single city varies from block to block. There is a combinatorial number of environments that exists when it comes to traffic networks. Accordingly, the approach of training for each scenario will not be practical moving forward. A potential solution to this challenge is to use transfer learning. Transfer learning is the use of knowledge gained from a source task to bias the learning process on a target task while forming a set of good hypotheses. A zero-shot transfer is where no learning is done on the target task and is analogous to out-of-distribution generalization in supervised learning. Kreidieh et al. (2018) investigates the transferability of knowledge from a circular source environment to an open street network environment and shows that knowledge transfer is possible between these two sources using zero-shot transfer. Ongoing research is looking into the possibility of learning from a single policy and applying the findings on many different scenarios.

Ultimately, the goal is to develop a set of techniques that can help analyze the mixed autonomy in the existing surface transportation system. This requires rigorous studying of mixed autonomy systems; DRL is a promising technique for the resulting modeling toolkit. However, there is a long way to go to build a toolkit for analyzing the whole urban system and translating the modeling results based on simulation to support real-world urban planning decision in different network architectures.

## 2.6 Traffic State Estimation with Physics Regularized Machine Learning: A New Insight into Machine Learning Applications in Traffic Flow Modeling[6]

Traffic state estimation (TSE) is the precursor of a variety of advanced traffic operational tasks. As the traffic sensors on freeway networks can only cover a limited range of areas, TSE is a useful tool to provide full-field traffic information. TSE models estimate flow rates and speeds over the whole network. Most TSE models in the literature are derived from macroscopic traffic flow models. In the early research stages, macroscopic traffic dynamics are found to be similar to hydrodynamics. The associated models are formulated for ideal conditions and significant effort is needed for their calibrations.

[6] By Terry Yang, University of Utah, U.S.A.

It is also difficult to work with the noisy and fluctuated data collected by traffic sensors. To deal with the noise and fluctuations, stochastic traffic flow models are used. These models may be divided into two types: stochastic extensions and stochastic formulations. Stochastic extension models add Gaussian noise to the model expression in order to quantify the noise from the sensor data. However, they can produce mean dynamics that do not coincide with the original deterministic dynamics due to non-linearity. Stochastic formulation models do not have this inconsistency problem; however, they might lose the ability to obtain a mathematical solution due to the lack of a closed form expression/methodology. With the increase in data availability, researchers in recent years started to look into more data-driven machine learning (ML) approaches. In general, the data-driven ML models can outperform the classical traffic flow models; however, the performance of these models still heavily relies on the quality and the quantity of available data. In order to mitigate such limitation, Physics Regularized Machine Learning (PRML) is introduced. PRLM is a novel modelling framework which encodes the classical traffic models ("physics models") into the ML framework: output from the physics models is used to later train the regularized ML model to improve the model performance. If compared to classical traffic flow models, PRML can effectively capture data uncertainty and reduce the efforts associated with model calibration. If compared to ML models, PRML is more robust as it better handles noisy training data (through a Gaussian process – GP) and is more explainable in terms of model performance.

Yuan et al. (2020) develop a stochastic physics regularized Gaussian process (PRGP) which uses a Bayesian inference algorithm to estimate the mean and the kernel of the PRGP. A physical "regulator" based on macroscopic traffic flow models is also developed to augment the estimates via a shadow GP and an enhanced latent force model is used to encode physical knowledge into stochastic processes. Based on the posterior regularization inference framework, an efficient stochastic optimization algorithm is developed to maximize the evidence lower bound of the system likelihood. The model is evaluated using four detector data sets from I-15 in Utah, US. The results from the case study show that all four PRGP models perform better than some standard ML models in terms of flow estimation and produce comparable and acceptable results in terms of speed and travel time estimation. Moreover, the findings show that encoding a bad physics model into the PRGP can downgrade the model performance. PRGP models also produce better estimates than the physics models. To study the robustness of the proposed PRGP model, artificial noise is added to the training dataset while keeping the test dataset constant. With noisy data, PRGP models produce acceptable estimates of flow, speed and travel time and the errors from this type models are less than the other ML and physics models. Adding more sensor data into the training dataset further improves the models' performance.

The PRGP can greatly outperform physics models in capturing the data uncertainty and fluctuations. When the training dataset is sufficient and accurate, PRML only slightly outperforms standard ML models in terms of speed, flow and travel time estimation. With noisy data, PRML is more robust than the ML and physics models. It can be noted that encoding a more advanced physics model can help the PRGP produce better estimates, while encoding an inapplicable physics model can downgrade the model performance.

## 3 Discussion

Advancements in the sensing and computational technologies have made the collection and the utilization of big data to solve real-world transportation problems more practical. In recent years, numerous research efforts have been seen in utilizing data-driven and AI methods for automated vehicles modelling and traffic operations. Such research efforts have led to the following question: what will be the role of core traffic scientists in the transportation sector and are traditional traffic flow and traffic operations models still relevant? Though applying AI approaches doesn't need the fundamental understanding of traffic science, traffic researchers need to well define the problems at hand. Defining the problems' characteristics and boundaries is a key to develop efficient and reliable AI models to solve traffic problems. The fundamental understanding of the mixed traffic flow dynamics and network infrastructure and control specifications is needed to adopt AI models especially when modeling AV systems and estimating traffic states. Complex neural network models are not always required and do not always perform better. In fact, in some cases (e.g., when developing network level optimum signal control strategy) simple supervised machine learning algorithms perform better.

The ultimate goal for the future transportation system is to make the whole transportation system fully automated with an increased level of safety, reliability, efficiency, and sustainability. Towards achieving such a goal, present researchers focus on individual autonomy, connected and automated systems and their network level performance. In line with such research directions, this breakout session presents six research efforts.

*The keys findings from the presentations and the subsequent discussion were:*

- Human interaction with AVs might play a significant role in AV system performance in the future and it should be incorporated in the AV system design.
- The transportation sector is one of the major energy consuming sectors in the US and connected AV systems can play a significant role to increase the energy efficiency and thus decrease energy consumption.
- Supervised learning can have great potential for AV control and traffic operations. Particularly, the traffic domain expertise can be used to design the problems faced, choose the proper AI or ML techniques to be adopted, evaluate the performance of the proposed methods, and interpret the corresponding results.
- Safety has been the upmost important factor in AV regulations and product developments. However, there needs to be a balance between safety and other performance aspects (e.g., congestion, energy consumption, fairness). Currently, safety remains the primary concern of the AV industry and the focus may be expanded to address other metrics/dimensions in the next 5 to10 years.
- There is a lack of available toolkits to analyze the impact of mixed autonomy. Understanding mixed autonomy can help make better policy to smoothen the transition process from manually operated traffic components to fully automated transportation system.

*The aforementioned findings motivate the following research needs/outcomes and may guide future research associated with adopting AI in AV modeling/development and traffic operations:*

- It is important to devise regulations/policies to guide the development and the adoption of AV and AI technologies (e.g., those related to safety and ethics). This is a challenging but a very important task with direct impact on the humanity.
- Two questions were raised regarding the role of data in adopting AI: (1) how much data is needed to evaluate the performance of a model? (2) How much data is needed/enough for AI training? The first question has been well studied. Although it is generally agreed that more data is beneficial for AI models' training/learning, the "enough" part of the second question has not been well addressed. Overall, more data collection and sharing efforts are needed.
- More collaborations are needed among government, academia, the AV industry, etc. For example, if researchers do not understand how AVs function, they will find difficulties in thoroughly evaluating the AVs' impacts on surface roadway network performance. Similarly, AI experts and transportation engineers should work closely to better address practical problems encountered on a daily basis in complex environments.
- Experts form academia and the industry need to connect with the policy makers to make informed decisions.
- Joint research between the government and the AV industry is needed to develop standards associated with insurance, security and communication strategies (e.g., vehicle to infrastructure -V2I- and vehicle to vehicle -V2V- communication standards).

**Acknowledgements.** The authors would like to acknowledge the remaining breakout session organizers: Alireza Talebpour (University of Illinois at Urbana Champaign), Mecit Cetin (Professor, Old Dominion University), Mehdi Hashemipour (Data Scientist, US Department of Transportation), Mo Zhao (Research Scientist, Virginia Transportation Research Council), Mohamed H. Zaki (Assistant Professor, University of Central Florida), Pan Lu (Associate Professor, North Dakota State University), Simeon Calvert (Delft University of Technology), Steven Mattingly (Professor, University of Texas at Arlington), and Xiaoyu (Sky) Guo (PhD Student, Texas A&M University – College Station). Additional thanks go to the TRB Committee Section/Committee chairs for their leadership and sponsorship of the session: Robert Bertini (Oregon State University), Soyoung Ahn (University of Wisconsin, Madison), and Sherif Ishak (Old Dominion University). This book was made possible with extensive outreach and synthesis efforts while coordinating the event details with the AVS2020 organizing committee.

## References

Abbink, D.A., Mulder, M., Boer, E.R.: Haptic shared control: smoothly shifting control authority? Cogn. Technol. Work **14**(1), 19–28 (2012)

Hao, P., Wu, G., Boriboonsomsin, K., Barth, M.J.: Developing a framework of eco-approach and departure application for actuated signal control. In: 2015 IEEE Intelligent Vehicles Symposium (IV), pp. 796–801. IEEE (2015)

Hao, P., Wei, Z., Bai, Z., Barth, M.J.: Developing an adaptive strategy for connected eco-driving under uncertain traffic and signal conditions. UC Davis: National Center for Sustainable Transportation (2020). https://doi.org/10.7922/G2F18WZ1

Kolekar, S., de Winter, J., Abbink, D.: Which parts of the road guide obstacle avoidance? Quantifying the driver's risk field. Appl. Ergon. **89**, 103196 (2020a)

Kolekar, S., de Winter, J., Abbink, D.: Human-like driving behaviour emerges from a risk-based driver model. Nat. Commun. **11**(1), 1–13 (2020b)

Kreidieh, A.R., Wu, C., Bayen, A.M.: Dissipating stop-and-go waves in closed and open networks via deep reinforcement learning. In: 2018 21st International Conference on Intelligent Transportation Systems (ITSC), pp. 1475–1480. IEEE (2018)

Laval, J., Zhou, H.: Congested urban networks tend to be insensitive to signal settings: implications for learning-based control. arXiv preprint arXiv:2008.10989 (2020)

Melman, T., De Winter, J.C.F., Abbink, D.A.: Does haptic steering guidance instigate speeding? A driving simulator study into causes and remedies. Accid. Anal. Prev. **98**, 372–387 (2017)

Melman, T., Beckers, N., Abbink, D.: Mitigating undesirable emergent behavior arising between driver and semi-automated vehicle. arXiv preprint arXiv:2006.16572 (2020)

Wu, C., Kreidieh, A., Parvate, K., Vinitsky, E., Bayen, A.M.: Flow: Architecture and benchmarking **for reinforcement learning in traffic control. arXiv preprint** arXiv:1710.05465, October 2017

**Yuan, Y., Yang, X.T., Zhang, Z., Zhe, S.: Macroscopic traffic flow modeling with physics regular-**ized Gaussian process: a new insight into machine learning applications. arXiv preprint arXiv:2002.02374 (2020)

# Autonomous Shuttles and Buses: From Demonstrations to Deployment

Katherine Turnbull[1](✉), Cynthia Jones[2], and Lily Elefteriadou[3]

[1] Texas A&M Transportation Institute University of Florida Transportation Institute, 3135 TAMU, College Station, TX, USA
k-turnbull@tti.tamu.edu

[2] DriveOhio, 1610 W. Broad Street, Columbus, OH, USA
Cynthia.jones@drive.ohio.gov

[3] University of Florida Transportation Institute, 365 Weil Hall, Box 116580, Gainesville, FL, USA
elfter@ce.ufl.edu

**Abstract.** This chapter presents information on automated shuttles and buses, which are being piloted, demonstrated, and deployed in downtown areas, university campuses, business parks, entertainment complexes, and other areas. The chapter focuses on the presentations and discussions at a breakout session at the 2020 Automated Vehicle Symposium (AVS). The session and this chapter highlight the experience planning, procuring, operating, and evaluating automated shuttles and buses to help inform future decision-making. Many projects were on hold in 2020 as a result of the pandemic, or pivoted to food delivery or other alternate uses. Areas for additional research and ongoing information sharing are also summarized.

**Keywords:** Autonomous shuttles · Driverless shuttles · Automated shuttles · Automated buses · Autonomous buses · Driverless buses

## 1 Introduction

Numerous pilots, demonstrations, and deployments of automated shuttles and buses continue in the United States and other countries. These services focus on enhancing mobility and accessibility on regular routes, providing first- and last-mile trips, and improving transportation options for individuals with disabilities. The AVS 2020 breakout group highlighted pilots and demonstrations in California, Ohio, Florida, Texas, and Minnesota. Projects in Scotland and the Netherlands were also discussed. Participants in the session shared experiences with autonomous shuttles and buses, highlighting lessons learned and tips for others interested in similar applications. Participants discussed best practices to help inform decision-making, identified research needs, and identified methods for ongoing information sharing.


G. Meyer and S. Beiker (Eds.): AVS 2020, LNMOB, pp. 73–80, 2022.
https://doi.org/10.1007/978-3-030-80063-5_7

## 2 Examples of Automated Shuttle and Bus Pilots, Demonstrations, and Deployments

### 2.1 Bishop Ranch Shared Autonomous Vehicle Pilot Program, Contra Costa County California

The Contra Costa Transportation Authority (CCTA) Shared Autonomous Vehicle (SAV) Program includes three phases. The project represents a partnership of the CCTA, GoMentum Station, Bishop Ranch, and EasyMile. Activities included procurement of the EasyMile vehicles and obtaining federal and state approval for operating the SAVs on public roads, initial testing of the SAVs at GoMentum Station, operating autonomous shuttles on public roads in Bishop Ranch, and deploying SAVs on routes serving Bay Area Rapid Transit and bus stations throughout the county.

The phased project highlighted the benefits of conducting testing in a controlled environment first before deploying automated shuttles on public roads. The experience also indicates the difference in operating speeds in a controlled situation and in a real-world environment, as well as the complexities of integrating a vehicle with traffic signals (vehicle to infrastructure [V2I]). Having a safety attendant on board was also noted as important.

Additional projects are also underway in the area. The Rossmoor First Mile/Last Mile Shared Autonomous Vehicles project in Walnut Creek focuses on increasing transit accessibility for the elderly. The County Hospital Accessibility Transportation project in Martinez includes on-demand wheelchair-accessible automated vehicle shuttle service to a public health facility to reduce missed appointments.

### 2.2 Linden Leap, Columbus, Ohio

The goal of the One Linden project, which is part of the Columbus Smart City, is to use self-driving shuttles to close transportation gaps in reaching public transit, affordable housing, healthy food, childcare, recreation, and education. The 2.8-mile loop connects St. Stephen's Community House, the Douglas Community Center, the Rosewind Resident Council, and the Linden Transit Center. One Linden fills a gap in transit service in the area, providing a first-mile/last-mile link.

The electric EasyMile vehicles can accommodate up to 15 passengers and are Americans with Disabilities Act (ADA) accessible. Safety operators are on board the vehicles. The vehicles were wrapped with "One Linden—Our Community, Our Future" to help with marketing and public education. The service began in late 2019 but was put on hold during the COVID-19 shelter-in-place requirement.

The test plan for Smart Columbus provides an example for use by others. The test plan included vetting the vehicles' capabilities over three phases, focusing on factory acceptance testing, preliminary acceptance testing, and final acceptance testing. A total of 115 tests were conducted over the three phases.

Outreach to emergency responders and the local community was also conducted as part of the project. This outreach allowed emergency responders to get hands-on training with the vehicles and to ask questions and obtain feedback. Holding an emergency responder tabletop exercise with seven scenarios was also part of the outreach effort, as was engaging with disabled individuals on the operation and use of the vehicle.

Ensuring adequate time for the National Highway Traffic Safety Administration (NHTSA) Federal Motor Vehicle Safety Standards (FMVSS) exemption was noted as a tip for others deploying similar projects. The One Linden project received closed-course approval November 7, 2019, and received full-route approval December 18, 2019. To address concerns over possible interaction with schoolchildren, service is not operated on weekdays from 8:00–9:30 a.m. and 2:00–3:30 p.m. The locations of some stations were also modified to help address these and other concerns.

Some of the lessons learned focused on geometric constraints, closed-course approval, and stakeholder coordination. There were some issues with geometric constraints, with the one driveway presenting problems for the vehicles. Obtaining closed-course approval helped make the full-route approval process go smoother. Stakeholder outreach and coordination was important for ongoing public support.

COVID-19 resulted in the suspension of service due to the stay-at-home order and social distancing. Longer-term social distancing guidelines have held off the restart of service. An alternate-use case for the vehicles was developed and implemented, with food pantry boxes being distributed into the community from St. Stephen's to the Rosewind Resident Council.

## 2.3 Texas A&M University SmartShuttle Demonstration, College Station, Texas

The SmartShuttle represented the coordinated efforts of the Texas A&M Transportation Institute (TTI) and Texas A&M University Transportation Services. A NAVYA autonomous shuttle operated on a 1-mile route on the main campus in College Station from September 9 to November 15, 2019.

The purpose of the project was to introduce students, faculty, staff, and community members to the technology. A student safety operator was always on the vehicle. A website provided operational status, route, and ride information. Outreach activities included class invitations, guest lectures, displays at the Fan Zone during football games, social media posts, the university president's video blog, and a ride with the university mascot, Reveille. Surveys were conducted with riders and non-riders, as well as with the student safety operators. Approximately 600 miles of service was safely operated, with 90% operating in autonomous mode.

Some of the lessons learned from the SmartShuttle included allowing adequate time for federal approval from NHTSA and for any needed local approvals. The importance of campus and community communications to potential riders was also noted. Recruiting, hiring, training, and scheduling the student safety operators took time and effort but proved to be a great experience for the students. Communication with NAVYA personnel occurred regularly before and during the demonstration, with weekly calls and as-needed texts. The use of a technical support team, including NAVYA and local personnel, was important for responding to issues.

Some issues were experienced with the vehicle during the demonstration related to air conditioning during hot and humid weather, the impact of radiated pavement heat on sensors, and operating in rainy and windy conditions. Other items to consider in operations include software updates, spare parts, towing needs, flat tires, and the need for repeat mapping after some software updates.

It is also important to consider all the other activities typically underway on a college campus in planning and operating an automated shuttle. Examples of these activities include football game set-ups, student resident move-in dates, campus events, and illegally parked cars. Staffing issues and new crosswalks also may need to be considered. An online survey was used and yielded good results.

The one key takeaway for others interested in deploying similar projects was the importance of first responder training. It was a critical element for preparedness, coordination, communication, and support during the demonstration.

## 2.4 I-STREET, Gainesville, Florida

I-STREET (Implementing Solutions from Transportation Research and Evaluating Emerging Technologies) provides a real-world connected and automated vehicle testbed in Gainesville focusing on mobility and safety, data analytics, and human factors. I-STREET is technology agnostic, takes advantage of the Florida regulatory environment, and focuses on developing best practices and processes. It further focuses on industry engagement and education, training, and certification. Additional start-up support is available through UF Innovate.

The Gainesville autonomous pilot project is a partnership with the University of Florida (UF) and the Florida Department of Transportation (FDOT). The City of Gainesville is also a partner in the project. FDOT is providing funding for the project. Transdev is the service provider, and EasyMile is the vehicle provider.

The autonomous shuttle route links the UF campus and downtown Gainesville. The project includes three phases. The first phase is from the downtown area to Innovation Square. The second phase extends the route to the UF campus. The third phase extends the route to Depot Park. Service on the initial phase began in February 2020 but was stopped in response to NHTSA directives. Services was resumed later in the year.

Florida I-STREET is also evaluating the Beep shuttle operating at Lake Nona. The service, using a NAVYA autonomous shuttle, launched in September 2019. The service was halted due to COVID 19 in March 2020. Service was relaunched in June 2020.

The evaluation of the Lake Nona shuttle includes collecting vehicle trajectories and design characteristics to assess the action of the AV shuttle while interacting with surrounding traffic. The evaluation will also monitor interactions and study behavior through videos, including in vehicle videos and videos at specific locations with and without the shuttle.

## 2.5 Texas Southern University Automated Vehicle Shuttle, Houston, Texas

The automated shuttle operated on a 1-mile closed-loop route on the Texas Southern University (TSU) campus in Houston, Texas. The pilot began in June 2019 and ended in February 2020. The Metropolitan Transit Authority of Harris County (Houston METRO) was the project lead. Project members included the Houston-Galveston Area Council (H-GAC) and TSU. The Texas Department of Transportation, the Texas Innovation Alliance, and the University of Houston also participated in the pilot.

First Transit Inc. was the shuttle operator, and EasyMile was the shuttle vehicle provider. An attendant was on board to take control if needed. The electric EasyMile 10 Gen 2 vehicle provided capacity for 12 passengers (6 seated and 6 standing) and provided access for passengers with reduced mobility. The Center for Transportation Training and Research at TSU conducted research related to vehicle operations, passenger engagement, workforce needs, environmental impacts, and safety.

The primary objectives of the pilot included gaining insights on the operational characteristics of the automated vehicle during different weather conditions and the electric battery capabilities under different temperatures. Gathering information on the perspectives of riders and vehicle attendants was another objective. Service was operated from 8:00 a.m. to 2:00 p.m. and 5:00 p.m. to 8:00 p.m. in the fall and spring. Service was provided from 8:00 a.m. to 4:00 p.m. during the summer.

The vehicle operated safely during the pilot. There were some issues with operation in the rain, but the sensors could be set to a higher rain level. The vehicle could accommodate wheelchair users, but it was not ADA compliant. With only one vehicle, there was no backup if there was a problem with the shuttle. There were also times when service had to be discontinued when onboard attendants were not available.

The vehicle electric charging capabilities were tracked to ensure that there was at least 20% battery life always remaining. Overall, the batteries performed as expected, even with the high temperatures and humidity of summer weather in Houston.

Survey results indicated that 43% of the respondents had heard a little about the driverless shuttle. A total of 33% of the respondents reported that the internet/blogs were their primary source of information. Ridership grew throughout the pilot, with the highest ridership in September, October, and November 2019.

Examples of lessons learned included understanding the potential operational limitations of the shuttle during inclement weather and knowledge of battery capabilities at different temperatures. Surveying riders to gain information on their perspectives on the service was also very beneficial.

Houston METRO has a long-standing record of embracing emerging technologies and innovative services. In 1997, METRO participated in the automated highway system with METRO buses platooning on the I-15 HOV lane in San Diego. In 2020, METRO was selected for the FTA's Accelerating Innovative Mobility (AIM) Challenge Grant. METRO's AIM Incubator project focuses on developing an automated electric shuttle bus using a Phoenix Motorcar Zeus 400 Motor F-450 Chassis, which is a mid-size vehicle. It is ADA, NHTSA, and Buy America compliant.

## 2.6 Rochester Automated Shuttle Pilot Project, Rochester, Minnesota

The project was selected through the Minnesota Department of Transportation (MnDOT) CAV Challenge request for proposal (RFP) process. It involves the operation of two EasyMile EZ10 vehicles in downtown Rochester, which is the home of the Mayo Clinic. There will be an onboard ambassador who can take over shuttle operation if needed. The service will be open to the public, with a minimum of 12 months of operation.

MnDOT is the project lead. The project partners include the City of Rochester, Mayo Clinic, and Destination Medical Center. First Transit and EasyMile are the two technology partners.

The project has four goals focusing on winter weather, infrastructure, public education, and mobility. The first goal is to advance the operation of AV technology in winter weather conditions. The second goal is to identify infrastructure gaps and solutions to the safe operation of AV technology on public roadways. The third goal is to engage and educate the public on the benefits and opportunities afforded by AV technology. The fourth goal is to enhance the transit experience for the citizens of Rochester and increase mobility in a high-demand downtown urban environment.

The City of Rochester also has goals for the project. These goals include enhancing the city's image as an innovative and forward-thinking city, applying lessons learned to future transit system technologies, and engaging with the public and gathering feedback on the use of the shuttles and AVs in general.

The proposed loop is approximately 1.5 miles long, circling the central part of the downtown area. The route includes stops at four locations on both sides of the street. Shuttle service is scheduled to begin in the summer of 2021. The pilot will operate for 12 months, with service provided from 9:00 a.m. to 3:00 p.m., 7 days a week. Anticipated infrastructure challenges include the connected signal systems in the downtown area, bus station designs, lane markings, and snow and vegetation management.

## 2.7 Arlington Driverless Shuttle Pilots, Arlington, Texas

The City of Arlington sponsored Milo, which was the first autonomous shuttle offered by a municipal government to the public on a continuous basis in the country. Milo operated from August 2017 to August 2018 on off-street trails in the Arlington Entertainment District. A trained operator was always on board and could take control of the vehicle if needed. Rides were free. The two major goals of Milo were to test the AV technology in a real-world environment and to educate the public on AVs.

Several lessons were learned from the Milo project. The pilot highlighted the importance of the deployment environment to successful operation of AVs. Ensuring a clear pathway for operations was key. Environmental modifications were necessary to address a few issues with the pathway for the Milo project. Vehicle procurement and insurance coverage were challenging and time consuming. The pilot also highlighted the importance of considering maintenance needs, including vehicle storage, cleaning, repairs, and transport to and from the operating site. Operation plans were prepared and used for operator training, public safety, and communications.

Milo served over 110 events, including 78 stadium events, 17 public demonstrations, and 18 special group tours. In a survey conducted with riders, approximately 90% strongly agreed that they enjoyed riding Milo, and 96% reported that they felt safe riding Milo.

The on-street operation of Drive.ai autonomous vehicles represented the second pilot conducted by the City of Arlington. This pilot program operated from October 2018 to May 2019. It included Drive.ai vehicles operating in mixed traffic at speeds up to 35 mph. Rides were free, and the service was open to the public. The Frive.ai pilot included 151 days of operation, from October 2018 to May 2019. A total of 755 AV trips were made, serving 1,424 passengers. A total of 451 AV passenger miles were driven.

The knowledge gained from the Milo pilot was very useful in developing the RFP for the second pilot. A turnkey operation was used for the second pilot, with Drive.ai responsible for all aspects of the service. The on-street pilot required fewer environmental modifications, and vehicles operated at higher speeds. Some frustration was exhibited from other drivers using the roadways, however, since the autonomous vehicles followed the traffic rules.

The city received $1.7 million from the Federal Transit Administration Integrated Mobility Innovation competitive grant program for Arlington Rideshare, Automation, and Payment Integration Demonstration (RAPID). The city is partnering with Via, May Mobility, and The University of Texas at Arlington (UTA). The project is integrating May Mobility AVs with Via's on-demand rideshare platform in the downtown area and on the UTA campus. The one-year demonstration, which is anticipated to begin in March 2021, will use five AVs, including one wheelchair-accessible vehicle. Free rides will be provided for UTA students.

## 2.8 AV Shuttles in the Netherlands

The HagaShuttle and the ESA Orbiter are two recent pilot projects in the Netherlands. The Future Mobility Network acted as the system integrator and project leader on both projects. The stakeholders included national and local governments, the road authority, vehicle manufacturers, infrastructure parties, and other groups.

The HagaShuttle operates at the Haga Hospital in the Hague. It serves hospital patients and visitors. Some of the patients have emotional or physical needs. The system connects a nearby bus stop to the main entrance via automated shuttle. A temporary route is in use due to ongoing construction, with a permanent route in the planning stage.

The European Space Agency (ESA) project ESA Orbiter operates on closed-off terrain, behind a gate, at ESA Estec. It serves primarily workers at the facility. The route is somewhat complex, serving parking areas and buildings. The vehicles operate at higher speeds (50 km/h), with multiple stops.

The HagaShuttle vehicle has performed well in mixed traffic situations and with pedestrians. In one year of service, there were only three out-of-path vehicle experiences. Passengers have been positive to the service.

The ESA Orbiter has also operated well. A clear line of sight is important, especially for shuttle sensors, given the complex route. The response from riders has been positive, but some note the low speeds as a limiting factor.

Lessons learned from the projects include maintaining flexibility since things often do not go as initially planned and innovation equals learning by doing. Cooperation among all groups is key, and maintaining flexibility is important. Further, developing stewards and sponsors is of great importance.

Additional challenges to scaling up deployment have also been identified. Examples of the challenges include increasing the vehicle's autonomy and capabilities to handle other traffic and more complex routes, increasing vehicle speed and capacity, and increasing operating capability in bad weather. Other challenges focus on increasing remote monitoring and decreasing the need for a steward in the vehicle.

### 2.9 CAVForth, Edinburgh Scotland

The goals of this project include developing a fleet of five automated buses and operating them in high-capacity service. The 14-mile route runs across the Forth Road Bridge and links the Ferrytoll park-and-ride facility with the Transport Hub at Edinburgh Park Station. Other goals are to optimize service by considering passenger needs and to explore the project's wider societal impacts.

The CAVForth project includes six partners. Fusion Processing is the project lead and the AV system provider. Stagecoach is the bus operator, and Alexander Dennis is the bus manufacturer. Transport Scotland is responsible for the route authority for the Scottish Government. BRL provides the simulator and cyber advice. Edinburgh Napier University is conducting the consumer and societal research.

The project includes launching a new commercial bus service and attracting commuters to use the service. The project also focuses on improving safety, fuel efficiency, and consistency of bus running times. The project provides firsthand experience with automated buses and will help influence the future of public transport.

The societal research focuses on three streams of work: empowering people to influence design of the service, monitoring feedback to improve the service, and building a broader understanding of the impact of automated buses. Over 500 people have engaged in discussions and design activities, including six workshops, a design jam, and a co-design panel.

The next steps in the project include completion of the safety case and approval by all partners and traffic authorities. On-route testing without passengers will be initiated, followed by full-service operation.

Some of the key lessons learned to date include the importance of including user requirements as the starting point for service design and the difficulty of reaching Level 4 operation. Taking a holistic approach that includes the onboard technology, bus operation, route monitoring, and customer input and feedback is also important.

## 3 Further Research

A number of areas for further research projects, pilots, and evaluation have been highlighted through projects, studies, and discussions at the AVS breakout sessions. Examples of topics for additional research include on-road and on-vehicle signing, sensor and battery robustness and performance, and remote supervision and monitoring. Other topics include common evaluation methodologies, core questions for user and public surveys and ensuring accessibility for all users. Continuing to share experiences with pilots, demonstrations, and deployments will also be important.

# Part III: Vehicle Systems and Technology Development

# Future Threats to Connected and Automated Vehicles

Jonathan Petit(✉) and William Whyte

Qualcomm Technologies Inc., San Diego, USA
{petit,wwhyte}@qti.qualcomm.com

**Abstract.** Automated Vehicles rely on intelligent systems to enable safe and efficient transportation. Thanks to robust perception and reliable communication, automated vehicles will reshape transportation services. However, the security of automated vehicles has to be guaranteed at the component level. In this chapter, we provide an overview of two threats to connected and automated vehicles (CAV), namely adversarial AI in perception system, and impact of Quantum Computer on CAV security.

**Keywords:** Post-quantum cryptography · Quantum computer · Automotive · Perception · Adversarial machine learning

## 1 Introduction

Connected and Automated Vehicles (CAVs) are complex systems of systems that have the potential to revolutionize transportation. As a component of a critical infrastructure, vehicles security is a key requirement. CAV security should cover sensors, processing units, controllers and network interfaces. Moreover, cryptography is holistically applied to provide confidentiality, integrity and authentication. In this chapter, we focus on robust perception system and post-quantum cryptography in the domain of CAV.

The perception system is responsible for capturing the scene, detecting objects, and associating tracks. Perception systems can use ultrasonic sensor, camera, radar, or lidar to sense the environment and rely on machine learning for object detection and classification [1]. In Sect. 2, we will describe the threats to perception system and techniques used to improve its robustness.

In Sect. 3, we will give an overview of the consequences the advent of a quantum computer would have on the automotive ecosystem. Finally, Sect. 4 concludes this paper.

## 2 Robust Perception for Automated Driving

Automated Vehicles equipped with sensors sense the environment by collecting sensor readings. These readings are then fed to a machine learning model in order to perform pattern recognition and object classification. For example, a camera output a grid of pixels that are fed to a neural network for bounding box detection. It is apparent that


G. Meyer and S. Beiker (Eds.): AVS 2020, LNMOB, pp. 83–91, 2022.
https://doi.org/10.1007/978-3-030-80063-5_8

feeding accurate grid of pixels is key to accurate detection in this example. An attacker can manipulate the object (or the sensor itself) in order to feed wrong data to the machine learning model. This type of attack is commonly called *adversarial example*. In [2] adversarial examples on vision-based perception system were demonstrated. Such adversarial examples can trigger a misclassification of traffic signs or sometimes worst, misdetection [3]. Figure 1 and Fig. 2 show examples of adversarial examples. In Fig. 1, the attacker performed a sticker attack on a stop sign. The carefully placed stickers mimic graffiti but effectively trigger misclassification by perception systems. In Fig. 2, the attacker crafter a real object such that 3D object detection algorithms do not detect it, which can be a serious safety hazard, especially when vehicles operate at high speed. Despite the effectiveness of adversarial examples, one should note that creating effective ones isn't a trivial task, and, as of late 2020, requires a high level of sophistication. Therefore, adversarial examples are sometimes considered a low risk in threat assessments. We refer the reader to the MITRE ATT&CK Adversarial ML threat matrix [4].

**Fig. 1.** Example of sticker attack that mimics graffiti and triggers misclassification [2]

**Fig. 2.** Example of adversarial object crafted to trigger misdetection [3]

In order to increase the robustness of machine learning against adversarial examples, one could apply the following techniques:

- Adversarial training: expose the system to synthetic attacks in order to cover edge cases,
- Generative models: estimate class-conditional likelihoods,
- Randomized smoothing: smooth decision boundary with noise.

As an example, [5] proposed to use adversarial autoencoders that give better likelihood estimates, and to train with discriminative loss to recognize out-of-distribution samples in order to defend against adversarial examples for a 10-class road sign dataset.

[6] presented an adversarial testing framework that identifies natural unsafe states found near safe ones in a simulated trajectory of learning-enabled controllers. Their

results demonstrated that the robustness of black-box reinforcement learning-enabled controllers can be significantly enhanced using learning-based detection and shielding approaches, without noticeable loss in performance.

But the two aforementioned techniques are examples of a large body of mitigations techniques. In [7], the authors provided a comprehensive survey on adversarial attacks on deep learning in computer vision. As of late 2020, the European Telecommunications Standards Institute (ETSI) has an Industry Specification Group dedicated to the development of guidance documents for Securing Artificial Intelligence. We recommend the interested reader to read the set of documents. From the sheer amount of mitigations techniques, it is apparent that identifying the right set of mitigations for CAV isn't an easy task. Especially, the community is lacking a gap analysis of mitigation techniques to understand the interoperability between them.

To conclude this section, one should remember that we don't have good defenses against adversarial examples, but some approaches are promising. In the meantime, the best practice is to keep machine learning models secret, be prepared to change them, and check consistency.

## 3 Connected and Automated Vehicle Security in a Post-quantum Computing World

### 3.1 Cryptography, Symmetric Algorithms, Asymmetric Algorithms, and Digital Signatures

Cryptography is a key enabling technology in the connected and automated vehicle space. The most familiar use of cryptography in popular culture is to provide confidentiality via encryption, in other words, to prevent unauthorized parties from reading other peoples' data. However, possibly even more important in machine-to-machine communications is cryptography's use as a building block of trust. Digital signature cryptographic algorithms, combined with digital certificates issued by a public key infrastructure, allow a receiver to have confidence that a received message has come from a valid sender, i.e. a sender who is capable of correctly making the observations of the world that go into creating the message and who can be trusted to create a truthful message. Digital signatures also allow for basic integrity of operation for each device, by enabling secure boot and secure update. The trustworthy operations and communications enabled by digital signatures are vital for Connected and Automated Vehicles (CAVs).

However, all the digital signature algorithms in widespread use can be rendered useless by quantum computers of sufficient size. Quantum computers are computers that use the entanglement between quantum objects to run certain specific algorithms faster than classical computers, which in contrast operate on standalone bits. Unfortunately, some of the algorithms that are speeded up most significantly – in some cases, exponentially – are the algorithms that break digital signature algorithms, i.e. that expose the private key from public information. If the private key is known, then anyone can generate signatures, and so a signature can no longer be trusted as coming from a specific, trustworthy device.

Although there are alternative signature algorithms that are believed to be resistant to quantum computers, the existing standard signature algorithms are well-established in product and in production processes. The industry therefore faces an organizational and supply-chain challenge in migrating to "post-quantum cryptography". Additionally, the post-quantum algorithms pose a technical challenge in some settings: existing "non-quantum-safe" algorithms can have very compact keys and signatures and fast operations, but the most attractive candidates for post-quantum use have large keys, or large signatures, or slow operations, or a combination of these drawbacks.

Quantum computers affect other security mechanisms, though less severely. For example, another form of cryptography uses *symmetric keys* rather than private and public keypairs to provide confidentiality and authentication. Symmetric keys require both sender and receiver to know the same key and, as such, systems based on symmetric key cryptography are more fragile and less scalable than systems based on public key cryptography. However, deployed systems exist that rely entirely on symmetric cryptography, or that use long-lived or hard-wired symmetric keys. Symmetric keys are reduced in strength by quantum computers, though not as much as public key operations: while popular public key algorithms are made effectively zero security by quantum computers, a symmetric key that takes an impractical $2^{128}$ classical operations to break is reduced to a more practical but still laborious $2^{64}$ quantum operations to break.

As such, this discussion of quantum computers and their effect on CAVs covers the following topics:

- What Needs to Be Changed in CAV Systems as a Result of Quantum Computers
- What is the likely timeline for this change
- What is the impact of this change on how systems are designed, manufactured, and used.

### 3.2 What Needs To Be Changed in CAV Systems as a Result of Quantum Computers

Table 1 lists current uses of cryptography in vehicle systems, both within individual vehicles and as used in vehicle communications, and the effect on these uses of quantum computing.

### 3.3 Quantum Computing Background and Timeline

Quantum computers are an experimental and evolving technology and as such it is unclear (at the time of writing, late 2020) when they will be available with enough computing power and enough accessibility that existing public key algorithms will be under threat. In fact, there is no expert consensus on this topic. The Global Risk Institute surveyed 22 experts on their estimate of the probability of full-scale quantum computing at a particular date. The results are shown in Fig. 3.

The earliest date by which 50% of the experts thought that there was a 50% probability of full-scale quantum computers being available was 15 years from the time of the survey, i.e. 2034. However, over 80% of the experts thought that the chance of a quantum computer being available 10 years from the time of the survey, i.e. 2029, was greater than 5%. Given the significant impact of successful development of a quantum computer, it seems wise for the CAV community to start planning immediately for a transition, in case it turns out to be necessary.

**Table 1.** Uses of cryptography in vehicles

| Activity | Type of cryptography | Effect of quantum computing |
|---|---|---|
| V2X | Public key for authentication/skey exchange, symmetric for data encryption | No incoming messages can be trusted |
| In-Vehicle Network | Public key for authentication/key exchange, symmetric for data encryption | Component authentication is forgeable; effect may be mitigated due to physical protection |
| Infotainment (Bluetooth) | Symmetric key derived from a PIN | Bluetooth security can be broken with max $2^{64}$ effort -- Not just an automotive problem |
| Tolling | Existing systems are primarily symmetric<br>Advanced systems propose integration with V2X and will use public key | Existing systems: longer keys, new tags for all<br>Advanced systems: public key, will be more complex to upgrade |
| Cloud service/local access | Public key for authentication/key exchange, symmetric for data encryption | Remote entities cannot in general authenticate/Exception: remote entities with a long-lived secure relationship based on a symmetric key |
| Software update | Public key in general | Software updates cannot be trusted in general/Exception: updates with security based on a symmetric key – this is not widespread |
| Secure boot | Boot loader – hash value burned in with fuses; Other components in the boot chain – public key for authentication | Boot loader can still be trusted, other components cannot |
| Data storage | Symmetric key stored in HSM | Data that was ever transmitted encrypted can be decrypted if intercepted; stored data encrypted with a public key can be decrypted easily if accessed |
| GNSS authentication | No authentication | Situation is no worse than now |

### 3.4 Impact of Transition

If quantum computers break existing public key algorithms, what might replace them? This question has been an active research topic in academic cryptography for decades, with numerous possible approaches having been proposed. What all proposals have in common is that they are based on different hard mathematical problems than the ones

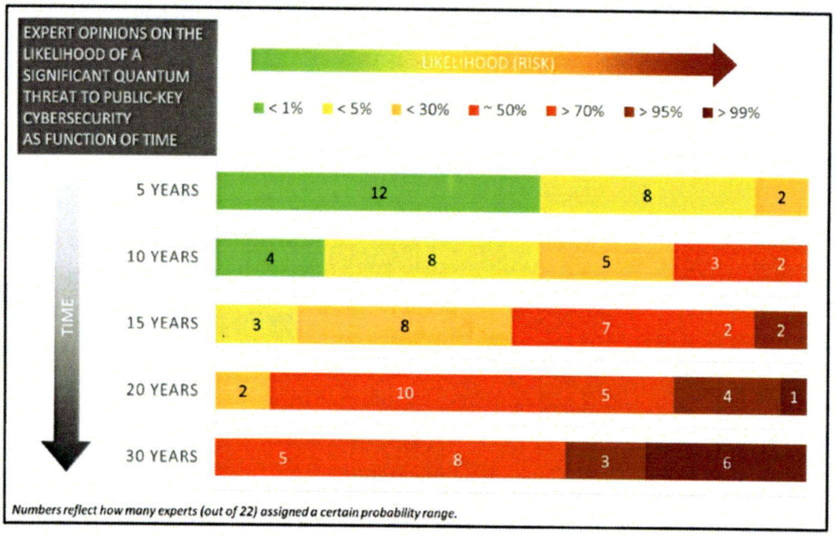


**Fig. 3.** Distribution of expert opinions as to when a full-scale quantum computer will become available (https://globalriskinstitute.org/publications/quantum-threat-timeline/)

that underlie the existing algorithms. Those problems are "integer factorization" and "elliptic curve discrete log"; candidate replacements are based on other problems, such as decoding a linear code, or finding a short vector in a lattice, or inverting an elliptic curve isogeny map. It should be noted that for these problems, all we can say is that no fast quantum algorithm to solve those problems is known *at the present time*. This means that for any candidate algorithm, all we can say for sure is that it is not currently known to be quantum-vulnerable.

As such, strategies for post quantum cryptography have tended to emphasize diversity, rather than selecting a single anointed algorithm for the world to move to. The most high-profile attempt to identify strong candidates is the competition run by the US National Institute of Standards and Technology (NIST), which is at the final stage as of late 2020 and expected to conclude in 2022. The competition has identified finalists based on three different hard problems (or, more precisely, three different sets of closely related hard problems) and intends to output one winner based on each of those problems.

Figure 4 below shows the signature size and public key size of the finalists. This is to be compared to a public key size of 256 bits/32 bytes and a signature size of 512 bits/64 bytes for ECDSA.

Clearly, no matter what algorithm or algorithms are selected, any operations that need signatures will incur considerably more communications overhead. (Processing power is not significantly higher, fortunately, although dedicated-silicon accelerators for a specific algorithm may become useless after a post-quantum transition). As a basic rule of thumb:

- Digital signature overhead goes from 100–300 bytes to 1000 bytes
- The size of a digital certificate goes from 200–500 bytes to 2000 bytes

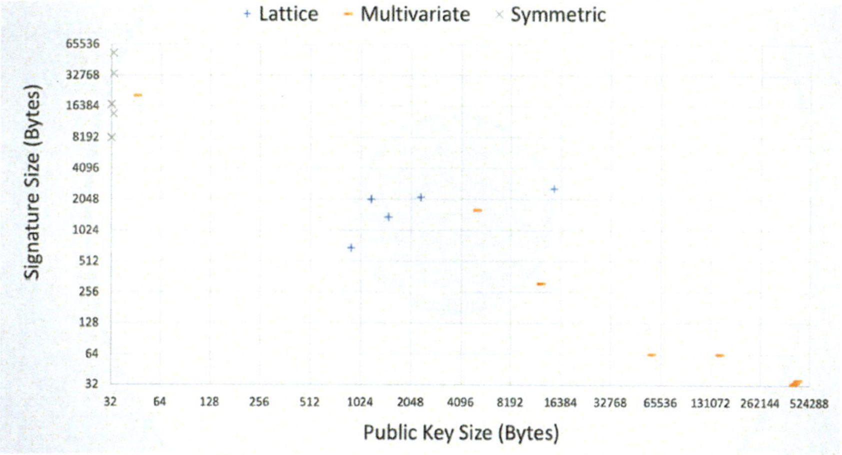


**Fig. 4.** Size of public key and signature for NIST competition finalists at the 128 bit (post-quantum) security level (https://csrc.nist.gov/CSRC/media/Presentations/Round-2-of-the-NIST-PQC-Competition-What-was-NIST/images-media/pqcrypto-may2019-moody.pdf)

Table 2 summarizes the effect of a post-quantum transition on the cryptography-using use cases identified in Table 1.

**Table 2.** Impact of post-quantum transition on automotive use cases

| Activity | Impact of transition |
|---|---|
| V2X | Need more channel capacity/fewer messages |
| In-vehicle network | Existing sensors/ECUs may be trustworthy enough in the context of wired communications, but communications that use public key cryptography need to be updated. Architectures should use gateway/firewall devices to prevent direct access to ECUs. Internal "cable size" should be increased to address bandwidth issues due to greater authentication overhead |
| Infotainment (Bluetooth) | Avoid sending sensitive information over Bluetooth |
| Tolling | Existing systems will need to update tags to use 256-bit symmetric cryptography. Advanced systems based on V2X will need to ensure that the system can accommodate the increased overhead due to increased size of cryptographic material |
| Cloud service/local access | PKI migration will be necessary. If a root of trust is embedded in hardware and uses vulnerable public key algorithms, the hardware will need to be replaced |

(*continued*)

**Table 2.** (*continued*)

| Activity | Impact of transition |
|---|---|
| Software update | Storage requirements increase due to larger signatures, but this is not significant in the context of the size of a typical software update |
| Secure boot | Boot loader – hash value burned in with fuses – does not change. Other components in the boot chain – public key for authentication – will need more storage for signatures |
| Data storage | Update any public key cryptographic mechanisms and consider how storage requirements are affected |
| GNSS authentication | No authentication so no impact; but it would be good if GNSS was better authenticated |

### 3.5 Conclusion

Clearly, the transition to post quantum cryptography will take time and careful planning. Although it is not clear when a move will become urgent, smart deployers will start making a plan sooner rather than later.

## 4 Conclusion

In this chapter we discussed two future threats of connected and automated vehicles: adversarial examples in ML-based perception system and quantum computer. Both threats are growing and require immediate attention from the CAV community. On a positive side, researchers have proposed promising solutions, but are not yet sufficient or comprehensive for CAVs.

Deployment of post-quantum cryptography affects the entire ecosystem and requires planning to be ready by the time quantum computers become a reality. Securing artificial intelligence raises research questions of explainable AI, models transferability, and role of hardware in securing AI.

Finally, we encourage the community to (i) investigate further the implications of quantum computers on CAV ecosystem, (ii) perform a gap analysis w.r.t. adversarial examples mitigation techniques, and (iii) not forget that the CAV ecosystem has many more attack surfaces that require equal attention.

## References

1. Marti, E., de Miguel, M.A., Garcia, F., Perez, J.: A review of sensor technologies for perception in automated driving. IEEE Intell. Transp. Syst. Mag. **11**(4), 94–108 (2019)
2. Eykholt, K., et al.: Robust physical-world attacks on deep learning visual classification. In: Proceedings of the IEEE Conference on Computer Vision and Pattern Recognition, pp. 1625–1634 (2018)
3. Cao, Y., et al.: Adversarial objects against lidar-based autonomous driving systems. arXiv preprint arXiv:1907.05418 (2019)

4. MITRE, "Adversarial ML Threat Matrix". https://github.com/mitre/advmlthreatmatrix
5. Ju, A., Wagner, D.: E-ABS: extending the analysis-by-synthesis robust classification model to more complex image domains. In: Proceedings of the 13th ACM Workshop on Artificial Intelligence and Security, pp. 25–36 (2020)
6. Xiong, Z., Eappen, J., Zhu, H., Jagannathan, S.: Robustness to Adversarial Attacks in Learning-Enabled Controllers. arXiv preprint arXiv:2006.06861 (2020)
7. Akhtar, N., Mian, A.: Threat of adversarial attacks on deep learning in computer vision: a survey. IEEE Access **6**, 14410–14430 (2018)

# Generic Cooperative Adaptive Cruise Control Architecture for Heterogeneous Strings of Vehicles

Carlos Flores[1], Xiao-Yun Lu[1,2](✉), John Spring[1], and Simeon Iliev[3]

[1] California Partners for Advanced Transportation Technology, University of California, Berkeley, S. 46th Street, Richmond, CA, USA
{carfloresp,xiaoyun.lu,jspring}@berkeley.edu

[2] Lawrence Berkeley National Laboratory, 1 Cyclotron Road, Berkeley, CA 94720, USA
xiaoyunlu@lbl.gov

[3] Argonne National Laboratory, 9700 S Cass Avenue, Lemont, IL, USA
iliev@anl.gov

**Abstract.** Cooperative Adaptive Cruise Control (CACC) systems uses wireless connectivity to guarantee string stability for platooning, which Adaptive Cruise Control (ACC) fails to provide. One of the constraints that hinders widespread adoption of CACC is that almost all developments and real validations have been done on strings of identical vehicles. This chapter summarizes the most recent efforts in the development of a generic control architecture that enables CACC on strings of vehicles of different makes/models/types, dynamics and powertrains. The developed hierarchical approach has demonstrated feasibility of CACC system on real vehicles, even at short time gaps. It is also robust in handling cut-in and cut-out maneuvers of other vehicles in public traffic.

**Keywords:** Cooperative Adaptive Cruise Control (CACC) · Passenger cars with different makes/models/types/powertrains · Generic CACC · Low-level speed tracking · High-Level Gap Regulation · $\mathscr{H}_\infty$ control · Linear Parameter Varying (LPV) · Actuation mapping · String stability

## 1 Introduction

Market penetration of vehicle automation has increased significantly in recent years. Systems like the automatic emergency braking, anti-lock braking and Adaptive Cruise Control (ACC) have demonstrated to be able to improve not only vehicle following behaviors and road safety, but also driving comfort. Despite the benefits that ACC systems have demonstrated and their large commercial presence, their impact on traffic flow and capacity has been demonstrated as negative [1]. On average, ACC systems set relatively large time gaps between vehicles due to cumulative delays from the downstream to the upstream in the vehicle following string, which would significantly reduce the traffic capacity [2]. Most importantly, it has been demonstrated that vehicles with commercially available ACC units do not satisfy the string stability criterion. Such condition requires


G. Meyer and S. Beiker (Eds.): AVS 2020, LNMOB, pp. 92–104, 2022.
https://doi.org/10.1007/978-3-030-80063-5_9

that the vehicle does not amplify the speed or acceleration disturbances coming from its preceding vehicle. In [3], it was demonstrated that a braking disturbance from the leader vehicle was amplified leading to risky maneuvering by the upstream vehicles in the following string.

To solve this problem, Vehicle-to-Vehicle (V2V) communication links have been introduced to Cooperative-ACC (CACC) system to remove the cumulative delays for a synchronized operation of all the vehicles in the platoon. Such systems have demonstrated to significantly increase the traffic flow and road capacity, by allowing short gap car-following with guaranteed string stability. However, recent studies have shown that real benefits of CACC adoption is mostly visible with high market penetration rates [4]. One of the major constraints that holds the widespread adoption of CACC systems is that almost all developed technologies have been designed and tested on strings of identical vehicles. This significantly constrains the formation of CACC strings of different vehicle types, makes and models on public roads, limiting the positive impact that this technology feasible.

Some recent works have been proposed to enable CACC between vehicles of different dynamics. The Grand Cooperative Driving Challenge gathered important European research institutions to accelerate and demonstrate CACC cross-platform feasibility. Other works have proposed solutions for CACC on heterogeneous strings [5–7], showing good results on simulation. Some other approaches have analyzed the implications of vehicles of different dynamics in the same string and how to guarantee bounded spacing error [8, 9]. Nevertheless, the performance limits of CACC on heterogeneous strings have not been explored in practice so far. In addition, more validation on real platforms is necessary to confirm the feasibility and for the development of guidelines of such technologies. Considerations for vehicles of different powertrains have not been derived yet in the literature, which is fundamental for nowadays increasing diversity of vehicles powertrain types in real-world.

In this work, a novel simplified CACC architecture is proposed to enable formation of strings of not only different dynamics but also different makes, models and powertrains. This architecture is generic in the sense that it is adaptable to any type of vehicles in a platoon with correct modelling and tuning. It also allows the driver to select a gap setting and a degree of performance. This architecture has been tested in real vehicle platforms with IC engine, hybrid-serial and hybrid-parallel powertrains, showing promising results even for short time gaps and other maneuvers such as cut-in and cut-out by other vehicles despite vehicles dynamics and powertrains heterogeneity.

## 2 A Generic CACC Architecture

The motivation of this paper is to propose an architecture that is generic to any type of vehicles for CACC or platooning. Its modularity is adaptable to different vehicle dynamics, powertrains and configurations, with proper modeling and control tuning. It takes advantage of the inter-vehicle communication network to augment the vehicle perception with variables of forward vehicles, significantly reducing the cumulative delays from the downstream vehicles to the upstream vehicles of the following string. The designed architecture is also able to perform CACC under different communication topologies—e.g. predecessor-only following, leader-predecessor following. It regulates the gap towards preceding vehicle, following a time gap based spacing policy. The time gap is defined as the bumper-to-bumper distance gap divided by the subject vehicle longitudinal speed. Three real platform vehicles are used in this work to test and validate the proposed architecture. These include an Internal Combustion Engine (ICE) Ford Taurus 2013, a parallel hybrid Toyota Prius Prime 2017 and a series hybrid Honda Accord 2014. The physical components of the developed framework are presented below.

### 2.1 Vehicle Setup

Different blocks of the CACC system implemented on the vehicle platform are outlined blow. The physical structure implemented on the three vehicles is depicted in Fig. 1.

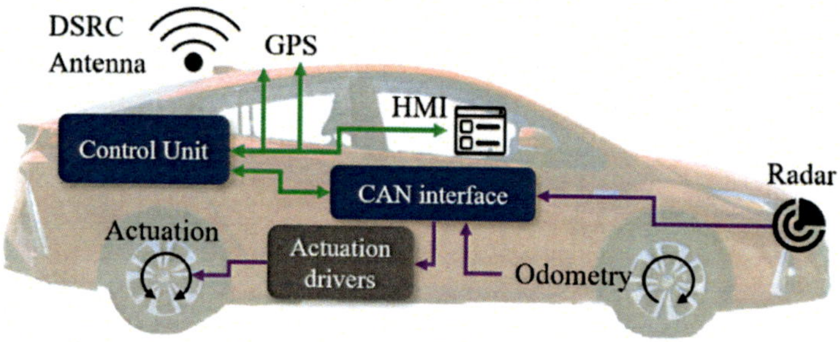


**Fig. 1.** Illustration of the physical architecture implemented on the vehicles.

A control computer (industrial PC-104) is installed on each vehicle that executes all the signal processing and control processes. A GPS is installed for synchronization and signals timestamping of all the vehicle in the string. A Human-Machine-Interface has also been developed and installed to provide the driver higher level control input and the display of relevant control information to the driver. Dedicated Short Range Communication (DSRC) Cohda wireless boxes have been installed, which broadcasts messages at a frequency of 50 Hz. All three vehicles are equipped with commercial ACC system, but they are quite different. They use radar for distance and relative speed estimation towards forward targets, encoders/odometry for vehicle speed measurement and a driver-by-wire network that reaches the actuators to control both braking and propulsion. In fact, the default ACC control was actually deactivated in the following sense: we only use the ACC activation to execute our CACC commands. The CACC control commands are sent to the CAN interface, which override the messages of the default or embedded ACC unit, by using a Man-in-the-Middle configuration. Regarding the braking, only the Honda Accord has the possibility to directly send the control messages to the brake actuator via an analog signal, whereas the other two vehicles communicate the reference deceleration to the vehicle ACC unit, which commands the brake actuator to track such signal.

The first car in the CACC string will always operate in ACC mode. We use our own designed and ACC instead of the default or built-in ACC.

### 2.2 Architecture Layout

For the generic control architecture, the same configuration is set on each vehicle. A hierarchical scheme is implemented, seeking a modular operation between the high and low-level control layers. The higher layer receives the subject vehicle state, time-gap and driving-mode (manual, ACC or CACC) settings by the driver, target perception and DSRC information, to regulate the relative speed and keep a safe stable distance. The key feedforward control algorithm here is to generate the reference trajectory for the subject vehicle to follow. The lower level control compares the proprioceptive sensors information against the given reference trajectory generated from the high-level, in order to generate the control command for the powertrain.

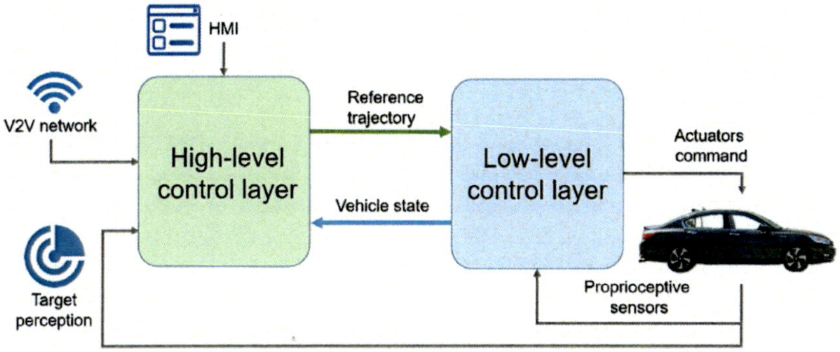


**Fig. 2.** Illustration of the physical architecture implemented on the vehicles.

For the CACC string configuration, the leader vehicle is set to track a desired cruise speed if there is no front target in certain range. If a target vehicle/object is detected, safe gap regulation is activated with the leader in ACC mode. The second vehicle, or first follower, implements CACC in a predecessor-only topology using the information of leader vehicle. The third vehicle receives the information from both leader and predecessor vehicles in a Leader-Predecessor-Following (LPF) topology, enriching the information available to perform an enhanced CACC following.

## 3 Low-Level Speed Tracking

The low-level control layer tracks the reference speed generated by the higher level. The design objectives of this structure are:

- Fast tracking of speed changes with robustness to disturbances.
- Closed-loop stability guaranteed with no overshoot.
- Consistent time and frequency response, despite the actuator types.

To tackle these design requirements, the following structure depicted in Fig. 3 is proposed for the reference speed tracking. The error between measured vehicle speed and received desired speed is processed by the robust controller in the feedback loop. The controller generates the desired acceleration, which is fed to an algorithm that estimates the ideal throttle or brake application level, based on the current vehicle speed and desired acceleration. This algorithm serves as accurate actuators' map that significantly enhances the translation from desired acceleration to actuation commands which may be different on vehicle types/makes/models.

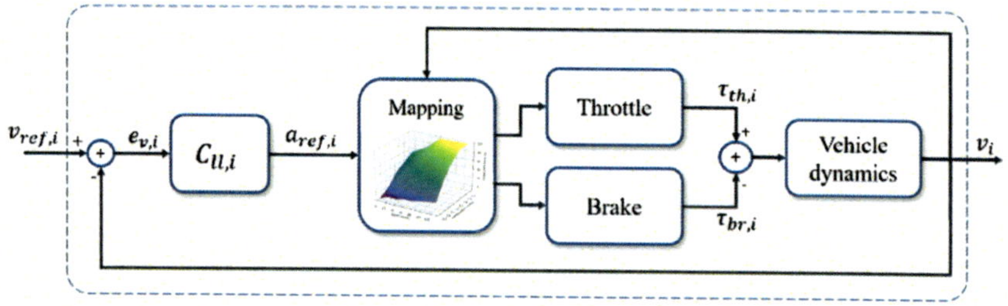


**Fig. 3.** Block diagram of low-level speed tracking architecture.

### 3.1 Actuator Maps Generation

This algorithm has been designed for the three vehicles used in this work. Firstly, the vehicles' longitudinal dynamics are estimated by performing coasting experiments from different speeds until stopping. This provides sufficient data to estimate the vehicle longitudinal dynamics parameters, such as inertia, aerodynamic dragging and rolling resistance coefficients [10]. After such models are identified, a dynamometer setup is loaded with the identified vehicle longitudinal model. Different tests are prepared with standard driving cycles that excite the powertrain dynamics for the entire speed and acceleration ranges. This yields extensive set of 3D points data in the acceleration vs. speed vs. throttle percentage level. All these points are linearly regressed to a surface function in the 3D space, providing a way to estimate the throttle application level that would yield a determined acceleration for any measured speed.

Regarding the brake actuator modelling, recall that only the Honda Accord platform brake is directly reachable. For this, different braking application levels are set starting from cruising at different speeds. This provides a deceleration profile that goes from the initial speed until stopping, for each braking application level. Subsequently, the speed vs. deceleration vs. brake application level points are linearly regressed to a function.

### 3.2 Control Synthesis

Once the vehicles' actuators have been modelled and mapped accordingly, the feedback controller is designed. A proportional controller has been used for its ease of implementation. Introducing additional lead or lag phase in the speed tracking loop is not necessary in this case, given the fast and accurate estimation of ideal actuation level provided by the developed mappings. The speed error is multiplied by the control gain to generate the reference acceleration sent to the maps. For each vehicle, the speed tracking gain is increased to get the highest response bandwidth within the set of gains that ensures no overshoot.

The low level control schemes are evaluated against a jerk steps-based speed profile. The speed tracking performance of the vehicles' low level blocks is evaluated in separate experiments with the same profile, which are illustrated in the same plot in Fig. 4.

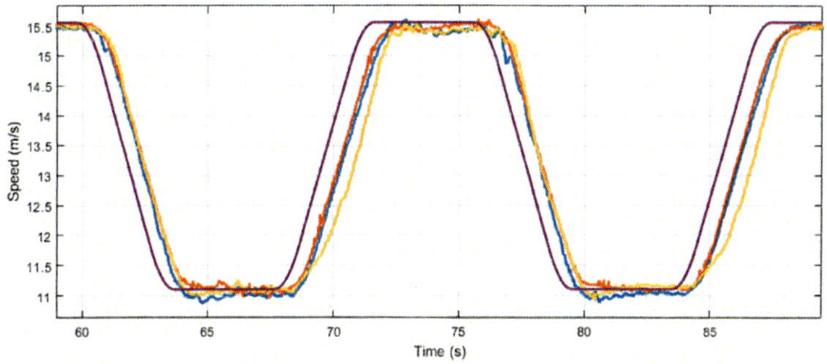


**Fig. 4.** Speed profile (purple line) and speed responses of Honda Accord, Toyota Prius and Ford Taurus (blue, red and yellow, respectively).

One can see that the no overshoot requirement is met, as well as the fast reference tracking. It can eb observed that acceleration and deceleration due to speed variations are tracked symmetrically. Only the Ford Taurus presents a slightly slower transient response for the acceleration response than for braking, given that ICE vehicles have a higher powertrain rotational inertia contributed by the transmission. Each vehicle low-level performance is finally enclosed in a second order transfer function that represent the low-level dynamics used in the higher level control design process.

## 4 High-Level Gap Regulation

The gap-regulation kinematics are handled by the high-level control layer. An illustration of the car-following architecture implemented on the three vehicles is shown in Fig. 5. The vehicles implement a speed decision algorithm, which takes the car-following speed and a desired cruise speed to generate the reference velocity to be tracked by the lower level controller. This architecture integrates the cruise control, ACC and CACC (only for non-leader followers) operation modes. The algorithm defines three operation regions: 1) no target or target farther than twice of the ACC time gap, 2) target vehicle within a region between one and twice the ACC time gap, and 3) target vehicle within the ACC time gap horizon. In the first region, the cruise speed is tracked. As the target vehicle approaches and enters the second region, the relative velocity is penalized, which guarantees a smooth transition towards the third region and car-following.

The car-following speed depends on the control mode that is activated—i.e. ACC or CACC. The first vehicle in a CACC string (leader) can only implement ACC, whereas the rest of vehicles can upgrade to CACC. For the ACC control loop, a feedback controller

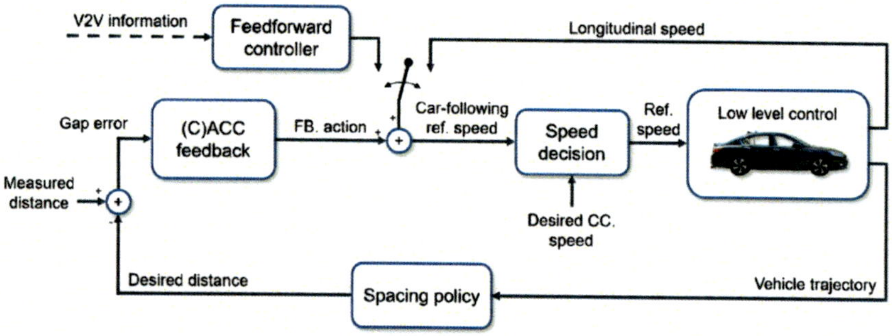


**Fig. 5.** Illustration of the high-level control architecture that enables cruise control, ACC and CACC operation.

processes the distance gap error and generates a feedback correction that is added to the vehicle current speed. The CACC control on the other hand, uses a feedforward controller to process the V2V information received from the connected forward vehicles beside the target distance and relative speed measured by radar. The output of such feedforward link is added to the feedback action to generate the car-following reference speed. The gap error is estimated comparing the measured distance gap towards the preceding vehicle with the desired distance from the constant time gap spacing policy block. The time gaps used on the ACC structure is larger than those for the CACC mode, given the enhanced tracking response of the latter. This structure permits to address the different design objectives in a modular fashion, with straightforward adaptability to any type of vehicle dynamics.

The transition task from ACC towards CACC control mode may be requested by the driver through HMI, ensuring at any time that the targeted preceding vehicle is available for CACC coupling. If the system evaluates that both target perception and V2V communication links are reliable, the transition is performed. Switching from CACC to ACC, on the other hand, is performed by the control system when a communication fault is continuously detected for a tolerance period exceeding the given threshold. All transitions (including those with the cruise control mode) are planned with bounded vehicle longitudinal jerk, acceleration and speed, which ensures a smooth and safe maneuvering.

## 4.1 Feedback Control Design

The modularity of the proposed CACC structure provides a two degrees-of-freedom design framework. The feedback system provides the required car-following stability and robustness against any possible uncertainty or disturbance in the low-level control system. The feedforward system enhances the reference tracking capabilities and increases the response bandwidth without affecting the loop stability. The structure design objectives are:

- Robustness to disturbances or uncertainties in the low-level control model.
- Guarantee of string stability with complementary sensitivity.

- Optimized control effort with bounded actuation bandwidth.
- The control performance vs. comfort tradeoff setpoint can be set by driver.

The design process is based on the parameterized model of the low-level speed tracking block. The feedback controller is designed as a Linear Parameter Varying (LPV) structure, where different controllers are designed for each operational point. The scheduling parameters are the target time gap and desired performance factor. Figure 6 depicts the LPV structure with its polytopes points. To get the structure output, the gap error is fed to all feedback controllers in the structure and linear interpolation is done between the vector of controller outputs, in function of the control settings selected by the driver via the HMI.

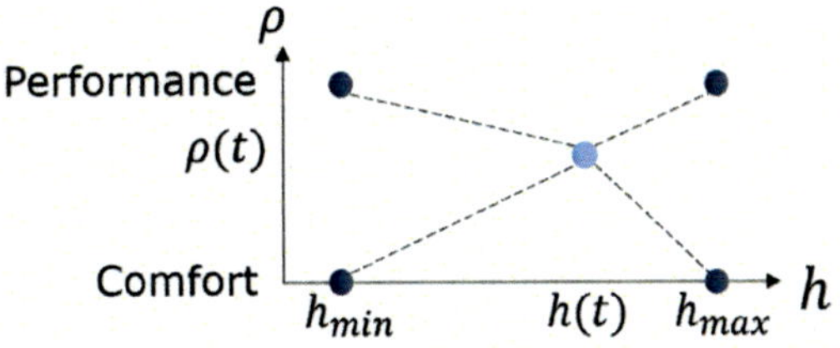


**Fig. 6.** Visualization of the control polytopes in the desired time gap vs. performance factor plane.

For each structure polytope, a feedback controller is designed by finding the optimal control frequency response that optimally shapes the structure functions: sensitivity, complementary sensitivity and control effort responses. The $\mathcal{H}_\infty$ control design framework is used, where template functions are plugged to the structure to define the desired shape of the closed loop functions.

- A good disturbance rejection at low frequencies is sought for the loop sensitivity, defined by the tracking error evolution in function of the structure input.
- The complementary sensitivity matches with the string stability function for this structure. The infinity-norm of this function is desired to be equal or less to guarantee string stability [11].
- The actuation response bandwidth is shaped as a low pass filter. This ensures that high frequency noise is not passed to the actuators and avoid possible saturation.

The values $h_{min}$ and $h_{max}$ are selected as the minimum and maximum time gaps available for the CACC control. Regarding the performance factor $\rho(t)$, the degree of performance of controller is manipulated through the shape of sensitivity and actuation response template functions. A larger bandwidth and less disturbance rejection are tolerated in exchange of higher tracking performance. The final controller for each polytope results as a lead compensator with cutoff, which gains, zeros and poles are optimally designed through this algorithm.

### 4.2 Feedforward Control Design

The CACC followers receive the reference speed of all connected vehicles in the string and use it to improve response with respect to disturbances from downstream vehicles. Recall that the first CACC follower uses a PF topology, whereas the second follower can perform LPF control. Figure 7 illustrates the feedforward structure blocks, where two stages can be distinguished. The first one adapts the reference speed signal applying a filter set with the ego-preceding or ego-leader vehicles' dynamics ratio, depending on the signal source. The second stage applies a low pass filter with a single pole in $\omega = -1/h$, where $h$ is the target time gap. This guarantees mathematically that as the communication delay converges to zero, the string stability condition will be satisfied for any time gap set [12].

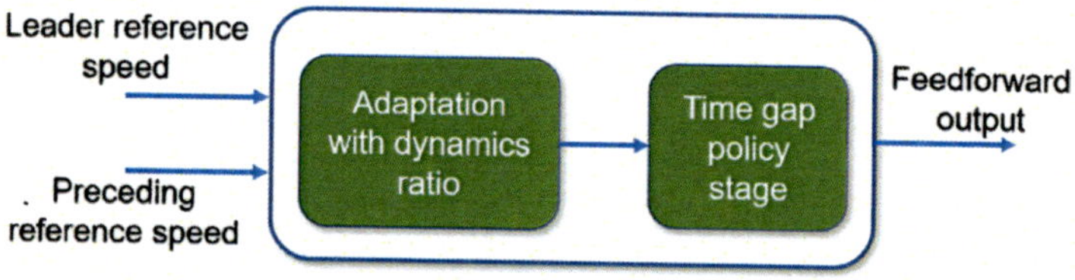


**Fig. 7.** Illustration of data flow and system stages of the feedforward block.

This method enforces that for time gap-based car-following, each string vehicle longitudinal trajectory should be a low pass filtered version of its preceding's [13]. The complexity that brings the string heterogeneity condition is then solved with the correct configuration of the dynamics ratio by either prior modeling of vehicles low-level dynamics or online identification algorithms, as proven in [14]. For CACC followers in third position and beyond, this process is executed for both the string leader and preceding vehicle signals, which are later linearly combined following the guidelines proposed in [9]. Vehicles with shorter response bandwidth is given more weight to leader vehicle's signal, whereas higher bandwidth vehicles are given priority to their preceding vehicle's signal.

## 5 Results Discussion

The generic architecture (Fig. 5) developed in this work has been tested in the vehicle platforms on different scenarios, including public highway and closed test tracks. Both ACC and CACC control has been tested on the leader and follower vehicles, respectively.

### 5.1 ACC Results

The testing for ACC structure has been set on the Toyota Prius platform and evaluated on highway I-80 in public traffic. Results are displayed in Fig. 9, where upper plot shows

**Fig. 8.** Photo of vehicle platforms on the closed testing tracks.

target and subject vehicle speeds. Lower left plot depicts the target and measured time-gap and lower right plot shows the spacing error. The controlled vehicle tracks very accurately all the speed variations from the target vehicle, which is manually driven. The control design, together with a sufficiently large time gap (1.3 s), demonstrate the string stability condition fulfillment. It is also visible that the speed propagation from target to subject is uniform and robust to different levels of acceleration and braking, which was one of the proposed design objectives. The time gap is seen to be accurately tracked and the spacing error deviates at most 3 m (at t = 500 s), which is acceptable given the nature of ACC systems and the relatively exigent deceleration from the target vehicle (around $-1.2$ m/s$^2$).

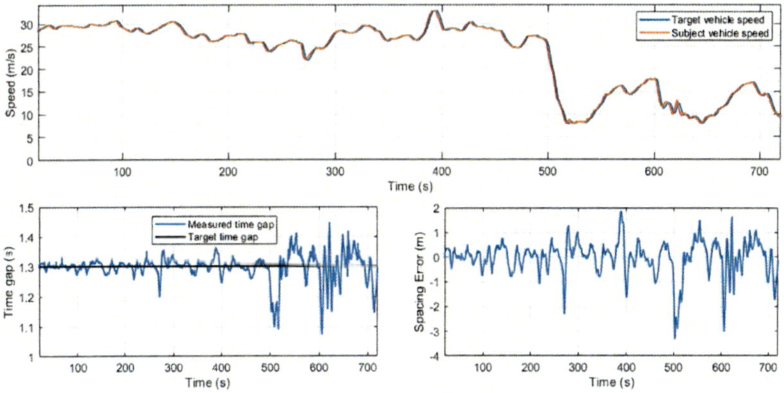


**Fig. 9.** Speed, time gap and spacing error results for the ACC system tested on highway I-80.

## 5.2 CACC Results

Regarding the CACC structure tests, several runs have been carried out to assess the performance of the system. Among the performed tests, there are highway runs on public highway, speed profile based on jerk steps and multisine speed profiles. Different time gap levels were tested to test the string stability limits and evaluate the potential impact on traffic capacity. Figure 10 depicts the multisine speed profile test on the closed test track.

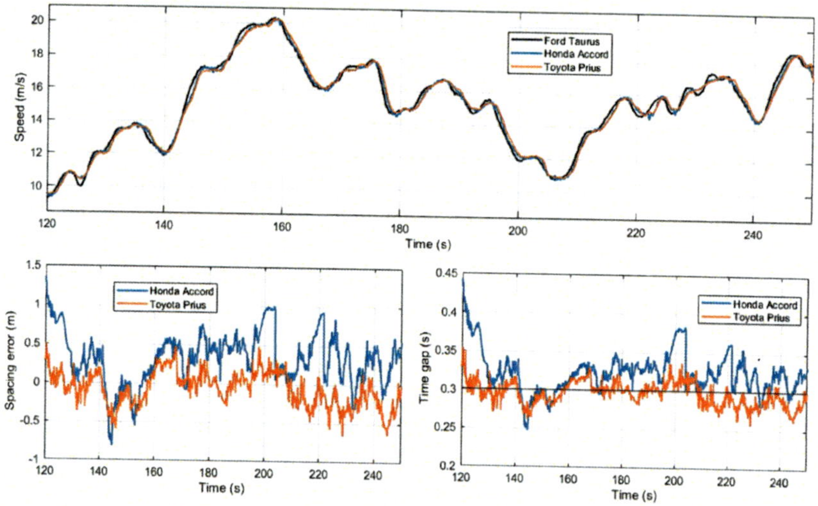


**Fig. 10.** Speed, time gap and spacing erroı results for the CACC system tested on Crows Landing airport facilities, Stanislaus County, CA, USA.

The profile designed for the leader vehicle comes from the sum of different sines of random phase, with frequencies selected within the spectral range that excites the string stability propagation function. The generated signal is set as the reference acceleration, which is then integrated to get the reference speed profile. The vehicle sequence is set as Ford Taurus, Honda Accord and Toyota Prius, in this order. As the leader vehicle follows the profile, one can see how the two CACC followers track accurately the disturbances generated by the multisine profile. Most importantly, the oscillations are not amplified from the downstream to the upstream, which demonstrates the string stability and its robustness despite the short time gap that is kept. It is also observed that the spacing error is greatly reduced compared to the ACC case, showing the significant enhancement that the feedforward controller provides to the reference tracking task thanks for the information passed by the V2V DSRC. Both follower vehicles show good tracking and damping of downstream disturbances, which indicts that the algorithm is potential to provide good performance despite the difference in vehicle types and powertrains.

## 6 Conclusions

A CACC architecture has been proposed that is generic to any type of vehicle dynamics and powertrain, which can be used for enabling CACC for strings of mixed vehicle types, makes and models. This is an important achievement towards popular adoption of CACC systems. It also represents progress towards mitigation of the negative traffic impact of current human drivers and commercially available ACC systems, which are not string stable in vehicle following and therefore could significantly reduce traffic capacity and safety with higher market penetration. It is noted that autonomous vehicles are among them since their longitudinal controls are ACC. The main research contributions of this work can be highlighted as follows:

- Development of a hierarchical architecture for CACC for car-following, where the low-level layer deals with longitudinal speed tracking based on nonlinear vehicle powertrain/drivetrain dynamics, while high-level regulates the time/distance gap in car-following based on vehicle kinematics.
- Each of the three low-level speed tracking systems performed very accurately, demonstrating that use of actuators mapping enhances the acceleration/deceleration tracking capabilities.
- A feedback control design with an LPV structure permits the selection of not only of the comfort/aggressiveness levels in function, but also the desired time-gaps (or the corresponding distance-gap).
- The inclusion of feedforward controller to process the V2V information permits to significantly increase the tracking speed and response bandwidth, without diminishing the loop stability due to the removal of the cumulative delay from the downstream to the upstream of the following string (platoon).
- Despite heterogeneous dynamics in the string, the use of leader-predecessor topology enhances the performance and response capabilities, leveraging the knowledge of all vehicle response dynamics.

It is also important to highlight that all developed algorithms have been tested and validated in highways and closed test tracks, in high speeds and stop-&-go traffic. It has been concluded that the best vehicle ordering should be based on increasing response dynamics bandwidth in upstream direction, avoiding actuators saturation in any of the CACC followers. It has been also confirmed that ICE vehicles account for slower acceleration response due to the engaged transmission, whereas hybrid vehicles can compensate such delay with direct electric-motor-driven wheels, at least for parallel hybrid powertrains. Future research could consider generating this approach to CACC/platooning of other vehicle types such as buses and heavy-duty trucks and other powertrains such as electric and hydrogen vehicles.

**Acknowledgments.** This research was supported by the U.S. Department of Energy (DOE) Vehicle Technologies Office (VTO) under the Systems and Modeling for Accelerated Research in Transportation (SMART) Mobility Laboratory Consortium, an initiative of the Energy Efficient Mobility Systems (EEMS) Program under the direction of Mr. David Anderson, with support of the Project Manager Erin Boyd and Danielle Chou who are gratefully acknowledged.

## References

1. Flynn, M.R., Kasimov, A.R., Nave, J.C., Rosales, R.R., Seibold, B.: Self-sustained nonlinear waves in traffic flow. Phys. Rev. E, **79**(5), 056113 (2009)
2. Wang, J., Rajamani, R.: The impact of adaptive cruise control systems on highway safety and traffic flow. Proceedings of the Institution of Mechanical Engineers, Part D: Journal of Automobile Engineering **218**(2), 111–130 (2004)
3. Gunter, G., et al.: Are commercially implemented adaptive cruise control systems string stable? IEEE Trans. Intell. Transp. Syst. (2020)
4. Liu, H., Kan, X.D., Shladover, S.E., Lu, X.Y., Ferlis, R.E.: Modeling impacts of cooperative adaptive cruise control on mixed traffic flow in multi-lane freeway facilities. Transp. Res. Part C: Emerg. Technol. **95**, 261–279 (2018)

5. Wang, C., Nijmeijer, H.: String stable heterogeneous vehicle platoon using cooperative adaptive cruise control. In: 2015 IEEE 18th International Conference on Intelligent Transportation Systems, pp. 1977–1982. IEEE, September 2015
6. Abou Harfouch, Y., Yuan, S., Baldi, S.: An adaptive switched control approach to heterogeneous platooning with intervehicle communication losses. IEEE Trans. Control of Network Syst. **5**(3), 1434–1444 (2017)
7. Al-Jhayyish, A.M., Schmidt, K.W.: Feedforward strategies for cooperative adaptive cruise control in heterogeneous vehicle strings. IEEE Trans. Intell. Transp. Syst. **19**(1), 113–122 (2017)
8. Shaw, E., Hedrick, J.K.: String stability analysis for heterogeneous vehicle strings. In: 2007 American control conference, pp. 3118–3125. IEEE, July 2007
9. Shaw, E., Hedrick, J.K.: Controller design for string stable heterogeneous vehicle strings. In: 2007 46th IEEE Conference on Decision and Control, pp. 2868–2875. IEEE, December 2007
10. Rajamani, R.: Vehicle dynamics and control. Springer Science & Business Media (2011)
11. Ploeg, J., Van De Wouw, N., Nijmeijer, H.: Lp string stability of cascaded systems: application to vehicle platooning. IEEE Trans. Control Syst. Technol. **22**(2), 786–793 (2013)
12. Naus, G.J., Vugts, R.P., Ploeg, J., van De Molengraft, M.J., Steinbuch, M.: String-stable CACC design and experimental validation: a frequency-domain approach. IEEE Trans. Veh. Technol. **59**(9), 4268–4279 (2010)
13. Ploeg, J., Shukla, D.P., van de Wouw, N., Nijmeijer, H.: Controller synthesis for string stability of vehicle platoons. IEEE Trans. Intell. Transp. Syst. **15**(2), 854–865 (2013)
14. Flores, C., Milanés, V., Nashashibi, F.: Online feedforward/feedback structure adaptation for heterogeneous cacc strings. In: 2018 Annual American Control Conference (ACC), pp. 49–55. IEEE, June 2018

# Part IV: Policy and Planning

# Public and Private Sector Collaboration to Advance Automated Driving Systems Testing and Deployment

Kristin White[1](✉), John Harding[2], Ted Bailey[3], Daniela Bremmer[3], and Robert Dingess[4]

[1] CAV Office, Minnesota DOT, Saint Paul, USA
Kristin.white@state.mn.us
[2] Connected and Automated Vehicles and Emerging Technologies, Federal Highway Administration, Olympia, USA
john.hardling@dot.gov
[3] Washington State DOT, Olympia, USA
{BaileyTe,BremmeD}@wsdot.wa.gov
[4] Mercer Strategic Alliance, Inc., Fredericksburg, USA
RDingess@mercerstrategic.com

**Abstract.** This Chapter is intended to lay the background for meaningful conversations between public sector and private industry, to understand the barriers and needs for automated driving systems (ADS) and AV testing and evaluation. This includes understanding the capabilities that need to be implemented both in the ADS and on the roadway to support safe and effective ADS operation and understanding what is needed to design and conduct joint tests or pilots. This session aims to identify a path forward to overcome specific, private and public sector barriers, facilitate, and advance AV testing and deployment.

## 1 Introduction

The Chapter sets the stage for the partnership discussion by first discussing, defining the ADS-Roadway testing, and evaluation continuum. The FHWA will discuss its current ADS-Roadway Test and Evaluation Framework project, which is intended to understand how developers want a roadway environment to support development of their ADS system; how States and local jurisdictions ensure ADS-vehicles deployments operate safely; how ADS perceives the roadway environment; and what investments jurisdictions, can make to support safe ADS operation. This session will also discuss the benefits of simulation, test track, and open road environments.

This Chapter pursues the following goals:

- Foster understanding of public and private sector needs to advance automation
- Build relationships and increase partnerships between the public and private sector to advance testing and deployments on public roads
- Identify possible methodologies and approaches that support collaborative testing


G. Meyer and S. Beiker (Eds.): AVS 2020, LNMOB, pp. 107–114, 2022.
https://doi.org/10.1007/978-3-030-80063-5_10

- Discuss use cases/applications that fill gaps in previous testing and deployments
- Identify tangible public and private sector public-private partnership opportunities to advance testing and deployment
- Leave session with focused next-steps to advance public-private AV/ADS deployment project(s)

We hope to replicate the successful previous interactive public-private partnership discussions from previous years in this virtual format. There is a strong need for partnership dialogues to understand how public and private organizations operate, understand our constraints and how we can together overcome those challenges to advance AV and ADS.

## 2 Testing and Development to Facilitate ADS-Roadway Integration

### 2.1 FHWA ADS-Roadway Collaborative Testing and Evaluation

- Theme of AV 3.0 is to understand the barriers and challenges to advance ADS testing.
- The ADS demonstration grants focused on the goals of safe integration of ADS, data sharing so support broad analysis, and collaboration.
- Stakeholder engagement and outreach strategies are critical to pilots' addition to technical data analysis. The FHWA provide some examples of work scopes and tasks that support strong ADS testing.
- Process needs to include the pre-testing development phase (the why/what), the test definition phase (the what and how), the test execution phase (the execution), and the test completion phase (project evaluation).

### 2.2 NHTSA AV TEST Initiative

- 1st platform connecting public, all levels of gov't, AV testers and industry to share information about testing
- Genesis was based on the lack of media and anecdotal information with a lack of centralized information which impacts public trust and confidence in AV safety
- Public-private stakeholders can provide accurate and efficient access to information
- Engagement goal is to create an open forum to exchange safety information, including public meetings, regional and town hall meetings. Sec. Chao notes that AV technology is not advanced enough to support wide-scale deployment, but some time they will be.
- On June 15th, NHTSA launched AV TEST to support this
- Consumer confidence in technology is important for successful deployment
- NHTSA currently working to create an online, public-facing platform that shares local, state and national data. Initial web pilot includes 10 companies – including Beep, Toyota, Navya, and EasyMile, and 9 states – CA, FL, PA, FL, TX, AZ, UT, GA. It is open to all.
- Q: Whole FHWA testing framework is designed to accommodate simulation testing, that decreases the use cases, then go on a test track, and then move into roadway testing.

# 3 Public-Private Partnerships to Establish Successful AV On-Road Testing Legislation and Requirements

States and private industry will discuss their priorities and challenges with advancing AV testing at the state level. Interactive polling will be used throughout this portion of the session. Presenters will discuss the different testing processes and will be a focused discussion about state-level testing to support ADS development, including:

- What is the range of testing requirement across the various state's legislation?
- What ADS testing is required for permit approval? Who conducts?
- What testing requirements present challenges to ADS developers?
- Are there are any monitoring requirements? What do States want to monitor?
- Are there inconsistencies among State requirements that make it difficult to establish multi-state development tests?
- What tests present challenges or may inhibit testing in a state?

## 3.1 Shuttle Deployments and Partnerships

- Patchwork approach needs to be addressed
- Minnesota's innovation program allows for innovative procurement
- EasyMile technology runs on a virtual rail; it is key to understand that there are over 230 deployments that are partnerships with agencies and end-users that had a vision for autonomy
- EasyMile is using autonomous technologies in COVID-19 to support food shelves and supply chain management. COVID-19 has pushed industry to pivot and innovate and NHTSA has quickly approved these efforts.

## 3.2 Pennsylvania

- Mark Kopko, Director, Office of Transformational Technology, Pennsylvania DOT/Melissa Froelich, Senior Manager, Government Relations, Aurora
- AV testing taken place in PA since 2011.
- PennDOT's approach allows them to have safety in mind while also being flexible. They can update their approach within days whereas the states that have regulations can take years to change
- In development of guidelines, they engaged industry; industry then informed the guidance. They focus on safety driver training and safety culture.
- This aligns with DVS/DOT expertise
- Aligns with US DOT's expectations of how state DOTs should be involved
- Addressed data privacy and data sharing needs and expectations
- Participants must submit a notice of testing. Need basic safety information, crash-reporting procedures, pre- and post-trip planning. Once approved, PennDOT sends a letter of authorization.
- Aurora prides themselves on open lines of communication. They have integrated the Aurora Driver on Chrysler Pacificas and are also advancing trucking systems. They do off-road virtual testing.

- Aurora appreciates the Pennsylvania Advisory Committee's feedback. Pennsylvania does not have a clear path to deployment. PennDOT's guidance for AV on-road testing is a collaborative process, which is a model for all.
- Important to understand that the state DOTs have critical expertise the federal government does not have.
- Patchwork approach is confusing for both the public and industry

## 3.3 Virginia

- Amanda Hamm, Connected and Automated Vehicles Program Manager, Virginia DOT/Anupam Malhotra, Director, Connected Vehicles & Data, Audi.
- Virginia has several test beds/testing facilities that can reproduce many weather environments. They are also focusing on rural Smart Roads. Virginia's VCC system supports cloud data processes, mobile devices, work zones, and monitoring systems.
- Audi focuses on vehicle-to-infrastructure (V2X) systems. Even with considerations "TraffficLight Information" launched in 216. 80% of cars have Audi Connect and it operates in 15,000 intersections nation-wide. It gives a countdown in the dashboard that indicates when a light turns green. Second is an advised driving speed for Eco-driving.
- Audi also focusing on work zone warning – received feedback with traffic information. If you advise driver that a light is going to turn red, their driving behavior changes. This behavior change is important to understand when moving from a comfort/convenience service to a safety service like work zones. Work zones are now communicated directly into the vehicle. Virginia DOT is facing challenging work zone safety issues that this technology helps solve.

## 3.4 Washington

- Driver Assistive Truck Platooning Ted Bailey, Cooperative Automated Transportation Program Manager, Washington DOT/Amanda F. Anderson, External Affairs Manager, Peloton.
- AV Testing Executive Order and the Governor's Cabinet focuses on how we can all be successful together.
- State established 5-year working group, including public-private subcommittees on licensing, safety, infrastructure, liability, health, and workforce
- Even after developing a framework, little testing was occurring. State DOTs want industry to engage them. For example, Peloton engaged the DOT directly and that helped pave the way to support testing and deployment.
- $5M liability minimum eliminated some testing partners
- Washington focusing on SAE Levels 4–5
- Peloton: SAE Level 1–2 system that supports human driver. PlatoonPro technology similar to adaptive cruise control, except that the system controls the gap in following distance. Still includes active safety features line lane keeping.
- Work closely with DOT staff to seek clarifications from advisory councils and regulatory bodies

### 3.5 Florida

- Raj Ponnaluri, Florida DOT, Connected Vehicles, Arterial Management, Managed Lanes, Florida DOT/Joe Moye, CEO, Beep.
- Important to take field data, and do data testing in-house to calibrate and validate data and determine if the models fit. You need to also develop new scenarios. The V2X data exchange.
- Florida has implemented a new statewide security credentialing management system statewide (SCMS) to test map data, basic safety messages. They use SAE J2735 BSM for testing.
- Also working with ATMA and CARMA
- Beep is focusing on how they deploy these shuttles in real-life situations, including intersections, to develop mature platforms. Beep believes safety is 1st and foremost
- First route was a single-mile in retail town center. In less than 5 months, they have taken 8,300 cars of the road by engaging the communicant
- 4 key takeaways:

  o Planning
  o Public-private partnerships
  o Collaboration
  o Advancing ADS technology as we learn

What is missing from the equation? A private sector partner perspective – Rob Dingness, Mercer Strategies – Need uniform standards, including traffic control standards, to ensure that we have safe, reliable systems.

## 4 Discussion

- What are some strategies to objectively test an AV system that may be dynamic over time (AI-based)?

  o Test environment: Take data and test them in a simulation environment; calibrate and validate the date, and re-apply in the field to verify the results.
  o Use real-time data platforms to collect, analyze and disseminate data fields comprising the APIs. 3. Utilize test beds in MI and Suntrax (FL) and elsewhere to try these out.

- Is there a plan to integrate the FHWA's testing plan with simulations? In other words, does it make sense to prove tests are safe before on-road testing?

  o FHWA's testing plan can accommodate real-world testing, simulations, and off-road testing. Need procedures that are flexible and can validate data

- What kind of metrics do you use to determine when you can safely deploy a shuttle and how do you measure if you are getting better?

- o Before any deployment, we ensure that we complete a Site Assessment Report - it is an approximately 200 page document that outlines all the risks within the route in question, and any mitigations we have. It is the basis of our safety management plan that is shared with NHTSA and is part of our ISO9001 certification. As for 'how are we getting better' - comparing what we used to deem a risk, today is something we mitigate with the advancement of the technology - i.e. unprotected left hand turns used to be a no-go. Today, we have the ability to scan the whole area to ensure it is safe to proceed in automated mode."

- If guidance is voluntary, what is the value to the AV companies of getting a certificate/testing permit? Do companies test without the certificate?

  - o All active testers in Pennsylvania have gone through the authorization process. In fact, our second authorized tester, Qualcomm, submitted their notice of testing prior to publicly announcing a presence in the Philadelphia region and beginning on-road testing. We take this as testament to our guidance having value and being reasonable. There are multiple reasons why testers comply with our guidance.

1) To demonstrate their commitment to safety as seen by testers such as Aurora.
2) Avoid the need for the Pennsylvania General Assembly to draft legislation.
3) Peer pressure. When companies such as Aurora, Aptiv, Argo AI, and Uber complete the authorization process, it puts pressure on other testers to follow suit. Also, the media in Pennsylvania are very engaged in the automated vehicle industry. If a tester does not go through the process, media will raise questions on how they can ensure the safe operations if they have not gone through the PennDOT authorization process when their peers have.

- "Will the Audi system still work with the new FCC 5.9 GhZ ruling?

  - o Yes we are working with the exact same C-V2X frequencies that the FCC has proposed to allow within the 5.9 GHz spectrum"

- Does your DOT have an agreement to receive AV test data that may help provide insights on operational analysis? Any data that can be shared from Audi on the "time to green application?

  - o On the data-sharing question, I can confirm that Audi and our service provider Traffic Technology Services (TTS) are collaborating with the DoTs to develop reporting that will help with operational analysis. This is now in prototype form and we are fine-tuning with DoT feedback. VDOT works with our partners on each project to determine data sharing needs. For instance, our Research Council is often asked to conduct research and analysis on data gathered during a pilot. It really depends on the type of pilot and the needs of the partners involved. As part of Pennsylvania testing guidance, testers must provide a semi-annual data report. The report includes operational and economic development information.

- What would the state DOT representatives like to see from developers/OEMs to know that they are safe to operate on the public roads?
  - o Collaboration
  - o Policy development
  - o Understanding that industry is coming from business perspective; we come from citizen's perspective; speak same language

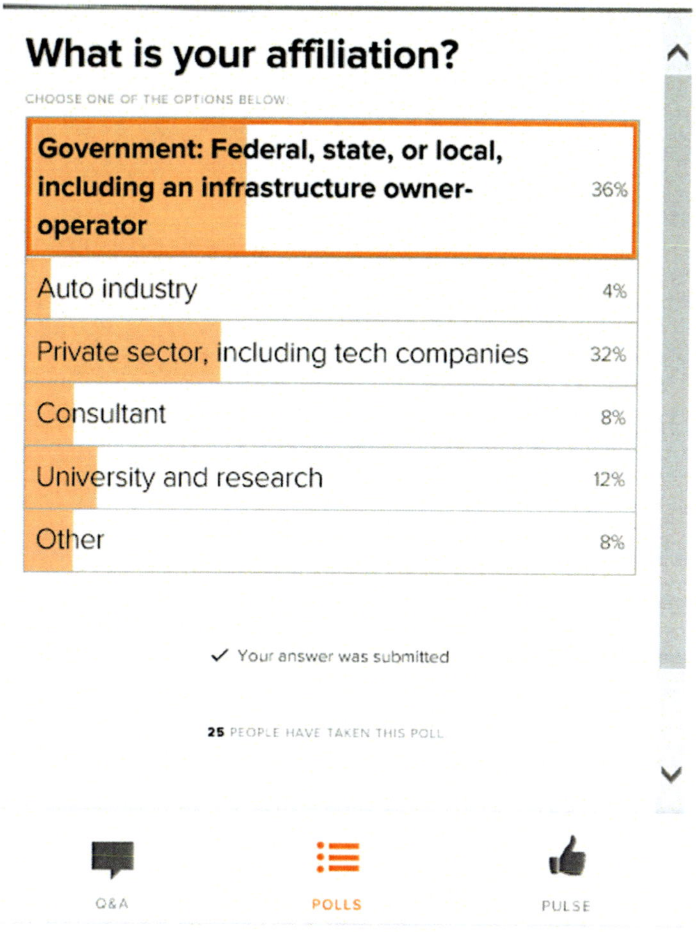


**Fig. 1.** Types of affiliation.

## 5 Conclusions

In order to advance ADS test, we need:

- Community and industry engagement

- Planning – have strong testing plans
- Public-private partnerships with transparent and open communications
- Ensure we're advancing the platforms as we learn
- Coordination across industry, e.g. AV TEST
- The importance of uniform policies and standards to avoid a patchwork approach (Fig. 1).

# Author Index





G. Meyer and S. Beiker (Eds.): AVS 2020, LNMOB, p. 115, 2022.
https://doi.org/10.1007/978-3-030-80063-5